THE FOURTH COYOTE:

A PERSONAL SEARCH FOR THE HIDDEN HISTORY OF EARTH

BY

KAREN DEGENHART

THUNDER STONE DREAM WOMAN

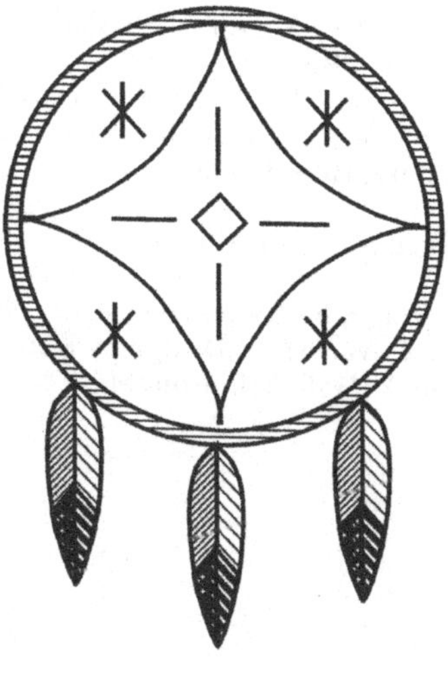

**THE FOURTH COYOTE:
A PERSONAL SEARCH FOR
THE HIDDEN HISTORY OF EARTH**

This book is dedicated to Robert Morning Sky, Elisabeth Kubler-Ross, and all my Spirit Helpers, from this or any other world.

ISBN: 0-916337-04-9, First Printing, Nov. 1995.
Second Printing, March 2010:
New, longer ISBN: 978-0-916337-04-9
Published by Thunderbird Publications/Koshare Press,
A Division of Karnak Enterprises, LLC.
480 Overlooked Way, Idaho Springs, CO. 80452.
(303) 520-4658, kdegenhart8@gmail.com

LIBRARY OF CONGRESS CATALOGING DATA:
Degenhart, Karen Sue: Born Jan. 9, 1954.
New Age, Autobiography, UFO, Extra-terrestrials, Native American Spirituality, Conspiracies, Lost Civilizations, Alternative, Speculative Ancient Earth History, Hopi Indians.

Printed in the United States of America.

Book layout and design, by Karen Degenhart. Photos and art work by Karen Degenhart. Cover Art or Design by Karen Degenhart. In other words, "I made this!" (With help from Mark Valco for the preparation for this printing.)

TABLE OF CONTENTS:

ABOUT THE COVER

Front Cover Photograph of Monument Valley, Arizona, by Karen Degenhart, 1988.
Back Cover Photograph of Karen Degenhart at Canyonlands, Utah by Don Devore, 1998.

Karen Degenhart, Fall 2009

CHAPTER ONE:

THE FOURTH COYOTE

The day I picked up my first book, *A Medicine Story*, from the printer, I went right over to Colorado's Psychic Center, and they bought the first two copies for their store. Then, I saw some small figurines of animals in onyx and other stone materials. I remembered one of my "Medicine dreams" in which I had seen fetishes such as these, and especially white onyx ones. In fact, the white fetishes were in the very dream which told me to do a Vision Quest. Now that my book was completed and printed up, I realized that I needed to "act out" more aspects of my dreams, in order to complete the Power available from them. So, I bought three small **coyote** fetishes, a white one, yellow one, and a brown one.

When I got home, and was setting them out on my dresser, I realized that I could have made a Medicine Wheel if I had gotten four coyotes! I should have gotten a red coyote, too," I thought. Then I'd have one for each Sacred Direction--Brown for West, White for South, Yellow for East, and Red for North." (Sometimes the White and Red are switched, depending on the Medicine Person. West is usually Black or Dark Blue, but Brown is also a dark color, so could be used in their place, I figured, as there were no Black coyotes at the store.) This was in January 1994.

I didn't go back to Colorado's Psychic Center until July 1994, the day I had turned in my *second* book to the printer! That day, they had a dish with items you could take for a **free gift**, and there was a **red coyote** in the dish, for the taking! I thought this was a strange coincidence, as I was going to buy one, anyway, to complete my Medicine Wheel. I had seen a coyote during my second Vision Quest, and it was one of my Totems.

When I turned in book number two, only six months after having printed my first one, the printer was surprised that I had done another one so fast. It actually took me a year to write the second book, but the typesetting went much faster, as I now had my secondhand computer to write on. All I had to do was hire someone with a laser printer to print the final manuscript from my disk. I also designed the cover, from a black and white photo of a painting I had done years before.

I told the printer, "Well, I'm not writing anything now." Yet, on my drive home that day, the words, "The Fourth Coyote" came into my mind, as the title of my next book!

When I picked up the new books, *Beautiful Simplicity: The Journey Into a Hidden Reality*, about my awakening to UFO's, ET information, and The Secret Government, I again went right to Colorado's Psychic Center, since they are just down the block from my printer. I was delighted when they again bought two of the first copies hot off the press.

Then a lady appeared, who worked there as a psychic reader. She saw my books on the counter, and expressed interest. I ended up talking with her, showing her my first book (still on the shelf), and told her it was about my awakening to Native American Spirituality, and leaving behind the mainstream church after that. She said that she was part Indian.

Then, she wanted to give me a brief "reading," as a gift. She asked my birthday, and tuned in. The first thing she said was, "Are you writing a third book?" I said, "Not yet, but I have been thinking about it. I was hoping to take a vacation from writing for a while!" She said, "You should do a third one." Something about her earnestness convinced me to start this third book, *The Fourth Coyote*, even though I really had no idea what it would be about.

Within a week, I came across some information that gave me a main insight for this book. It has to do with a re-writing of history on this planet! After I got this idea, I remembered that my Vision Dream in Boulder had said at the tail end of it, "If you like history, there is a lot that hasn't been written." The Vision Dream seemed to be alluding to a Hidden History of the Native Americans, but it goes beyond just that. There is a Hidden History of the Planet, and some Native Peoples have stories and artifacts that can provide us clues. Suddenly, and unexpectedly, I was off again on another journey of discovery. August 5, 1994.

CHAPTER TWO:

GOING FARTHER "OUT ON A LIMB"

I turned in the manuscript and cover illustration for *Beautiful Simplicity* to the printer on July 1st, 1994. On that day I got the fourth coyote fetish, a red one, and the idea for the title of this book.

That evening, I rented the video of Shirley MacLaine's metaphysical classic, **OUT ON A LIMB**. The last time I had seen it was in 1987, when I was in my final year of seminary, and it was first on TV. At that time, I was trying to be academic and mainstream, but this movie opened a crack in my thought. Perhaps it was "OK" to be metaphysical, believe in past lives, and UFO's--as well as to go beyond what the traditional church teaches. For some reason, I had suppressed these aspects of my own inner spiritual journey during the years after my Christian conversion, and while I was attending a seminary. This was a seminary which, even though it taught a radical modern theology of liberation, and allowed the student to come up with their own final theological interpretation of Christ, still occasionally made joking remarks about crystals and "New Agers" in the classroom. Of course, this was 1987--who knows what's popular among seminarians today, in 1994-'95?

I did in fact write a paper on Gnosticism, and one on the early Church Father Origin's view of reincarnation--and probably got A's on both papers for my scholarly approach. Yet, I still felt uneasy really sticking up for these beliefs in public there.

I remembered that there was something about UFO's in **OUT ON A LIMB**, and now that I had just published a book on that topic, I wanted to view the video again from my new perspective now seven years later.

I found it very interesting that one of the channeled readings Shirley had had, I think the one by Kevin Reyerson, told her that her British lover had been a diplomat with **extraterrestrials** during the time of Atlantis! I'm sure that did not seem as plausible to me when I first saw this video as it does now. In fact, his socialist leanings--taken to an extreme-- could even contribute to the "New World Order," (or **One World Government**, the evil one, which could easily give the bad aliens world control--not to mention what normal evil or control-oriented humans could do without the help of aliens!)

Her lover had a hard time believing in God or other spiritual concepts, too. Somewhat related to this, I had a strange experience the day before my UFO book came back from the printer, when I was at The Tattered Cover Bookstore here in Denver, feeling high at seeing my own *A Medicine Story* book for sale on the shelf there. I got into a brief conversation with a man standing there, and he looked over my book. He started to talk about how he didn't really believe in God, just in "getting on with life." Yet, he said something like, "The Universe is too big for there to be One Being in control of everything." He seemed to be responding to the mention in my book that I had once wanted to be a Christian minister. I said I thought it was probably only a matter of theological interpretation (as I know from my seminary training.) Then we parted ways, to look at other books.

I went to the UFO section next. Then, the weird thing was, just as he was about to leave the store, he came up to me again, and said, "The answer to your question, is 'Yes, UFO's are real.'" His eyes twinkled in an unusual way, and his remark seemed to be more than a joke, but a **confirmation** of fact. Then he left. I almost suspected that maybe he was some kind of alien, himself. And, if so, we can see, there are aliens who do not like the concept of God that we have here on Earth. I don't know if there is any connection between the strange man I met, and the atheistic beliefs of Shirley MacLaine's lover, but for some reason the connection between these two characters, and their trouble believing in God, struck me.

However, the Pleiadians are considered to be the Good Guys of the aliens by many metaphysical-type UFO buffs. (At least some Pleiadians are.) In Shirley's account, a Pleiadian woman had influenced Shirley's friend David to give her the three stones that she had put on a necklace. Later in the movie, when they were in Peru, David's Pleiadian woman friend, Mayan, telepathically guided him to find Shirley when her jeep was stranded -- and then guided him back, in total darkness, with no marked roads to follow. That was quite a feat--in the dark, I have even been lost on back roads between Idaho Springs and Central City! (And that was near houses and civilization!) August 10, 1994.

But, now I am seven years past 1987, when I and many others

have become open to the possibility of extraterrestrial contact. There is even the Fox TV show "The X-Files," which brings this whole thing out to the public for anyone to examine. The collective consciousness has moved forward. There have been so many sightings now, world wide, that most intelligent, open-minded people are probably ready and eager for some public announcement to be made, that **something** is "out there," --and for the Government to give up their need for any further cover-up of this information.

It was this year, 1994, that Dr. John Mack, the psychiatrist who has worked with "Alien Abduction" victims, and written a book about his cases, was on TV several times. We are being primed by the media to be open to the possibility of contact with aliens or extraterrestrials. We know by now (especially those who have read my book, *Beautiful Simplicity: The Journey Into a Hidden Reality*,) that nothing like that gets in the media unless it's being **allowed** out on purpose.

I've read in various places that an announcement that we have made contact with extraterrestrial life is supposed to take place in 1995 or 1996! As I write this chapter in September 1994, I hope that this book is done by then, **or** that I am able to chronicle these events during the course of my story.

Keep in mind, there are two points of view of this 1995 ET contact announcement. One is that it will be a wonderful revelation, a historic moment! The other is that it is being "staged" as the lead-in to the creation of an Evil One World Government! Thus, as remarkable as such an announcement will seem, we will still not be told the entire Truth!

One purpose for my writing this book at all is to reveal **information sources** that are difficult for the average person to come across, especially through regular media channels. To give more perspective, I tell **how** I came across this information. Thus, you can see how much time, effort, and "chance" play in the way I discover the things I write about. It is a lengthy learning process, and one never knows exactly where it will lead.

In my journey, I started attending a group called, "Global Sciences," which is a catch-all name for the avant-garde, cutting-edge, thought in politics, science, spirituality, and health. I first heard of this group from announcements at the Denver UFO Society meetings. Later, a man who attended the Koda Maka Dakota Spirituality classes, given by Grey Owl, told me he was a member of Global Sciences, and urged me to attend sometime. When I moved back to the city to take an apartment manager position in '93-'94, I started attending the Global Sciences meetings, since they were not too far from where I lived.

At one Global Sciences meeting in Spring of 1994, someone shared how great this magazine called, *NEXUS*, was! It was started in Australia, but has subscription offices in many countries, and now also in the USA, in Illinois. I looked at the copy they had that day, and knew I had to subscribe.

This *NEXUS* magazine has all the underground information one could want (or not want.): UFO, Secret Government, conspiracy information, medical cover-ups, plus book reviews, and ads selling very unusual videos. It's a fascinating, informative source for news about the sinister, Shadowy, side of just about everything that is being covered-up, especially in the United States. It is international, so people across the world probably know more about what's going on **here**--on the Shadow side-- than Americans do!

NEXUS isn't paying me for this advertisement--but it is well worth the $25.00 a year subscription to have this information. Write to: *NEXUS*, P.O. Box 177, Kempton, Illinois, 60946-0177, USA. This is the kind of information that will change your entire way of seeing reality, and politics.

I find it interesting that *NEXUS*, comes out of Australia! Australia is also supposed to be where the New World Government is going to be headquartered! (According to George C. Andrews in his book, *Extra-Terrestrial Friends and Foes*, the New Parliament Building in Canberra, Australia, is large enough to be used for a New World Government.) (Pg. 215, 216.) I have family who have immigrated to Australia, and a half-brother who actually works for the Australian Government now. I have been there twice so far to visit them. So many connections.... (September 18, 1994.)

Through *NEXUS Magazine* I discovered another important source for alternative information, and this is worth your trouble to get, as well. Adventures Unlimited, 303 Main Street, Post Box 74, Kempton, Illinois 60946. They may ask a dollar for the first catalog. This is where I first heard of David Hatcher Childress, a writer on "unorthodox" archeology and travel books. There will be more mention of his work later in this book.

The final source I will mention in this chapter is **Illuminet**

Press, P.O. Box 2808, Lilburn, Georgia 30226. From this source I discovered books such as *Casebook on Alternative 3*, by Jim Keith, and *Extra-Terrestrial Friends and Foes*, by George C. Andrews. They also put out the famous *The Gemstone File*, book, about Mafia Connections, and the JFK Assassination. There is some significant information in that book.

More of this later on. Time to switch gears. 10-6-94

Please note: Jim Keith has already died by 2010, so this address may no longer be functional. Some of his books can still be found at www.AdventuresUnlimited
Press.com.

CHAPTER THREE:

THREE JOURNEYS

For the Summer of 1994, I had a chance to take three camping trips. As I write this chapter, I am on the third one, sitting at a picnic table, by a rushing mountain stream, near Mt. Princeton, on the way to St. Elmo.

The first trip I took was to attend the Sun Dance, lead by David Swallow, Jr., at Buffalo Creek Campground, near Pine, CO. It is a very beautiful drive getting there. I pitched my tent there, instead of sleeping in my camper truck as I often do, and stayed three of the four nights.

For some strange reason, I have noticed a pattern of my boyfriend, Don, getting more emotionally confusing to me just before I go to the Sun Dance! Just before I left, we had a fight, and I wondered if I should consider not living with him, anymore. So, while I stayed at the Sun Dance Sunday and Monday nights, I went home Tuesday to take a shower, and attend Don's Nietzche discussion group. I also hoped to be able to mend any bad feelings between us. Somehow, this effort helped, and I returned to the rest of the Sun Dance feeling more assured that we could work things out. Don was getting worked up because he wanted to quit his pizza waiter job to spend all his time perfecting his new building technology ideas, but he couldn't quit yet, since he needed the money. The Summer was now half over, and he hadn't completed as much as he had hoped.

I'm glad to report that Don really became more considerate and relaxed once he finally quit the pizza waiter job sometime in August. Because his Rammed Earth Home is paid for, he can live quite cheaply on some savings for a few months, and even afford to travel to places like New Zealand or Australia for part of the rugged Colorado Winter.

I was lucky to get a small amount of unemployment from mid May through most of September. I had lost my HUD Resident Apartment Manager job because of being too zealous in my attempts to get enough evidence to evict the mentally ill residents who were actively using and selling drugs. They would prop open the security door, and there would be traffic day and night, disturbing other residents, who were even too frightened to call the police! So, I would call the police, for them, and my good deed led directly to my being fired, as somehow one cop found a "crack pipe" and gave it to me for "evidence." The day after I gave the "evidence" to my Property Manager Supervisor, I was fired! I later heard that the Mental Health Corporation of Denver wanted to contest our evictions of the drug dealers, who were scaring, and keeping awake all night, the other residents.

Some of the residents involved were just being used by the drug dealers, and had lost control of their own apartments, anyway. I could sit in my office and see groups of people all "visiting" the same apartment during the broad daylight, and yet the police could not do anything to catch them!

One time, I was helping out the under cover cops, and they just took a chance on catching some action. They called and told me they were on their way, and for me to buzz them into the building. By chance, the residents opened the door, and the cops found a warm crack pipe, and someone flushing some white stuff in a bag down the toilet!

After the guy who was caught was evicted, then things got worse. The traffic just moved to other apartments. The detective who had helped me get that "evidence" was put on some other kinds of cases, and my supervisors were obviously being pressured by the Mental Health people to stop trying to evict the residents. Anyone evicted from HUD Housing for drug use can never get it again, and these people had to live on SSI checks of around $400. a month. Where would they go? Sure, that would be a problem, but the ones being used by the dealers already could hardly sleep in their own beds; it was no mercy to allow them to stay in this situation, either. And the whole building was in turmoil because of it.

So you see, once again I tried to do the right thing in a corrupt system, and lost my job because of it. My resume is filled with such bizarre incidents, such as the time I was fired because I called a psychiatrist to have a woman hospitalized who was constantly cutting her wrist with tin can lids, and I was the only staff on a 26

½ hour shift, and I had to watch a houseful of other clients, as well, plus get some sleep during the night. My decisions always seem to rub someone the wrong way.

The day I was fired from the HUD Apartment Manager job, I was forced to move out of my apartment, supposedly in "three days," (though it took longer, because of all the stuff I had there on a third floor apartment with no elevator.) Luckily, I had Don to move back in with, but it was also an hour drive each way to bring truck loads of my things to his place.

The management company who hired and fired me also got out of paying any unemployment benefits to me because the building itself was supposedly my "employer," and the building was a small non-profit corporation! It was strange that the same group that owned this building also had some other buildings for the mentally ill, but they were incorporated separately. The rents for the studio apartments in my building were outrageously high for that area of Denver: $674. a month for a studio apartment! Maybe you'd pay that in Evergreen, but this was unusually high. However, you see, the management company would get a percentage of the monies they took in, so even though I got a free apartment and a final salary of $525 a month for working more than my supposed part time hours, the rent for the other apartments was guaranteed by HUD, and part of this Federal funding went to the management company, and the rest went to the utilities and the owners. Even with high utilities, there had to be some pretty good, stable profit being made by this "non-profit" building. Even empty apartments would be paid 85% of the rent by the Government, to the owners.

However, I lucked out, and got some unemployment, because a former job I had was in the right time span for that income to count. I had prayed with my Pipe for a good "vacation" during the Summer, so I guess I got an even longer one than I expected!

At the Sun Dance, I noticed an Eagle flying, and prayed to it, asking, "Eagle, please drop me a feather as a gift." The Eagle was not near where I was, and I knew it would be a miracle to find an Eagle feather, anyway. Then, I forgot my prayer.

A while later, the same day, I was sitting by my tent reading *Extra-Terrestrial Friends and Foes,* (by George C. Andrews, IllumiNet Press, GA.) when a lady came up to me, and gave me a small Eagle feather!

She said, **"Spirit told me to give you this Wamblei feather."** It was actually three very small feathers still attached together. My jaw dropped in surprise. I told her, "I prayed for an Eagle to drop me a feather just a while ago!" So, of course, my prayers for healing in my relationship also came true.

As I hadn't had a "real" vacation yet this Summer, by mid August I decided to take a four day camping trip during the Full Moon to get the wanderlust out of my system. Don could not take time off from his work on the concrete building panels, so I decided to go by myself. I enjoy taking some trips by myself, to meditate and reflect.

This trip ended up being more of an expensive chore than a vacation, as I ended up driving long hours through dry, barren land in Northwest Colorado, on hot days, without air-conditioning in my vehicle. It was still an adventure, however.

The most memorable time was the first night, camping at Dinosaur National Monument, Utah, after driving for eight hours. The moon was full, and the landscape was primeval. I felt a strong, peaceful energy at that campground.

I also enjoyed the lecture at the dinosaur dig sight. Here I learned that there was a "new" dinosaur that had been discovered, which was a meat-eater, and yet was found in a layer of earth much older than the plant eater found previously in the same area.

As the lecturer was speaking about the discovery of this dinosaur around 1990-1991, I suddenly remembered how the Native American Tony Shearer had phoned me, maybe in 1989, while on his way to Dinosaur National Monument to see some new discovery there. I thought it might be the same discovery, but that he could have heard of it before it was announced to the public. Tony Shearer always seemed to know about archeological discoveries that few people knew about. He was mentioned in my first book, *A Medicine Story*, as an expert on the Mayan Calendar, and the original person who pin-pointed the Harmonic Convergence dates. In that book, he was telling me about ancient sea fossils he had seen in Utah, in the Canyonlands area. From being there myself, I could see why Tony would be fascinated by Dinosaur National Monument.

On this trip, I went to "Blazing" Gorge, but it was too hot, and I was too worn out from all the driving, so I didn't do much there but visit the Visitor's Center, and photograph the Gorge. A video about the construction of Flaming Gorge Dam was quite interesting. President Kennedy was in the video, joking about not liking to push buttons, since they could lead to a nuclear war, as he was asked to push the button to get the dam's turbines started.

Then, in September, I decided to take one more trip, because my Unemployment had run out, and I knew I would be starting some work soon. This would be the "third journey" I mentioned in the chapter title. The last week of September was gorgeous, sunny, and the Fall colors were extremely powerful in mid and Southern Colorado during this week.

On September 26, I camped the first night of my trip at a gorgeous spot just off 285 South, the road on the way to St. Elmo. The National Forest Campground was FREE this time of year, and I found a spot next to a rushing mountain stream, near Mt. Princeton. The trees were all yellow, and the towering mountains surrounding the area were powerful. I took out my lantern, got on my jacket, and wrote for a while, sitting in the dark at the picnic table. It wasn't long before I began to get nervous of the sounds of the night, as a sign was posted warning campers of bears. Then I moved to the back of the truck, and wrote some more by candle light (as it's dangerous to use propane lanterns in closed areas, and I eventually closed the truck up to keep warmer.) The moon did not come out until later, so it was very dark, and I had to go to bed around 9 p.m. due to the dark and the cold, and nothing else to do.

Don couldn't come camping with me again because he had an important appointment with a licensed Engineer friend who was helping him to get the concrete panel idea approved as a marketable building technology. This is Don's best hope now of becoming self-employed, and maybe even wealthy, so that he doesn't have to go back to being a 43 year old pizza waiter with an engineering degree. As I was leaving, he hugged me close and promised to be able to come along next time I took a trip. That was sweet.

This is the life context around which my research into UFO's, Secret Technology, and the Hidden Side of Ancient History is occurring. As I discover more Secrets, you, the reader, will discover them with me. (9/26/94)

Still near St. Elmo, I woke up the next morning, thinking strongly about how some Indians may respond to the parts of my book *Beautiful Simplicity: The Journey into a Hidden Reality*, which mention them. Surely, they can see that I am philosophically "on their side."

I then drove to the ghost town of St. Elmo--an **extremely beautiful** drive because of the perfectly blue sky and bright gold aspen trees everywhere.

On my way to Pagosa Springs, I stopped at a little metaphysical and used book store called, **And Books, Too**, in Del Norte. The lady there encouraged me to have more faith in myself, and in my books. She said she would put up a flyer in the store, and would try to order my books later, once she got caught up from just moving there from Mexico. She does psychic readings, and makes medicine bags and dream catchers. She mentioned that the San Luis Valley was a "Sacred Valley." She reminded me of the people I met at the **Light Within** store in Estes Park, Colorado, who encouraged me to get my books listed with New Leaf Distributing Company, the distributor many metaphysical book stores buy from.

It took me until 4:30 p.m. to get to Pagosa Springs, Colorado, where the group that puts out the "MASH Unit," for awakening Extra-terrestrials, and the author of *ET 101: The Cosmic Instruction Manual*, are now located. Finding out more about this group was what drew me to Pagosa Springs.

Since I hadn't had a shower since I had left Idaho Springs, I paid $4.00 to swim in a Hot Springs pool. Then I had to back track a few miles to find an isolated National Forest area to camp in. As it was getting dark, I missed the regular campground, and instead camped up Turkey Creek.

Each night I camped on this trip, it got spookier and spookier. This time, I was in an isolated area that wasn't even a campground, all by myself! I had wanted to heat up something to eat, but instead, it was getting dark and cold so fast when I got there, that I just changed into my sleeping sweats, and snacked on some graham crackers and peanut butter, in the back of the truck. I made notes by candle light for this chapter, and went to bed by 9:00 p.m. just to keep warm, and hidden in my sleeping bag.

Pagosa Springs is not far from the Dulce, New Mexico area, near where an entrance to a secret underground military base is said to be located. This is an area where UFOs are seen, and cattle mutilations found. There were cattle not far from my camping spot, too.

As I was writing, a moth suddenly killed itself by diving into the candle wax! Isn't it strange that moths are attracted to that which kills them?

When I'm camping alone, I feel safer in my truck than in a tent, as there are wild animals, and even cows, often around. I keep the food locked in the front of the truck. And, I am happier if I don't see any people, as out here one never knows what they are up to. One man in a truck drove by both ways, apparently also seeking a campground, but I got the only spot that had a bit of an indention from the road, enough to pull in and camp with some privacy. So, now I prayed for safety during the night. I had my Sacred Pipe with me, which hopefully was providing some spiritual protection, but hadn't even had time to smoke it, yet this trip.

If there were UFOs here, I wouldn't even see them, hiding under covers in my truck camper top. My Hopi/Mexican friend said he had seen some UFOs a couple weeks previously, near Trinidad, Colorado. They were like moving stars, and did some maneuvers together before darting off in their unusual (non-conventional air craft) way. He believes they were "ours," and possibly from "Area-51." "Distance isn't a factor," he said.

I heard some coyotes yelping as I was getting ready for bed that night. The coyote is one of my power animals. The moon came out later in the night, but it was no longer full.

The next morning, I went back to Pagosa Springs and phoned the **"MISSION CONTROL"** people from a convenience store. Zoaz answered, and I was able to set up a meeting with some of them at a near-by coffee shop, since they were about to go there, anyway. I ended up talking mostly with a woman who now calls herself, **The Gaella Elan Rah**, which was something like her Inter-dimensional name. It turned out that she was formerly a social worker from Denver, and she had even worked for one of the agencies that later became the Mental Health Corporation of Denver. She use to frequent the same metaphysical bookstores that I know of there.

I decided to buy a T-shirt from them, since they spent time with me, and it would be an unusual souvenir. I later went to her home, and saw the base of operations for selling their books and tapes out of her garage. One whole wall of the garage was nothing but boxes of books containing, *ET 101: The Cosmic Instruction Manual.* I was told that this book was bought by a large publisher, Harper-Collins, and that these last few copies of the original self-published edition would be collectors' items. The author was Diana Luppi, now going by Zoaz.

The **Mission Control** group also offers a kind of spiritual counseling that goes by such names as "Device Removal and Activation, Light Body Alignment, Multi-incarnational Alignment, Genetic Code Adjustment, and Free Form Sessions." There are some suggested donations for this, as, in tiny lettering the brochure they sent me also says, "The Mission Control Mash Unit is an outreach project of Universal Life Church, Inc." I did not try these sessions out, primarily due to just barely having enough money to buy food on my trip home, while charging all my gasoline, but I congratulate this group for being creative and different, and even humorous, in their presentation literature, books, and services. I also met a woman who was planning on designing specialized and individualized clothing for those who want to dress for their inter-dimensional identity. They will soon have a store front and possibly be offering retreats and workshops. For further information, you may write to **MISSION CONTROL**, P.O. Box 2066, Pagosa Springs, CO. 81147. (In 2010, this may not longer be accurate.)

I would like to give the book *ET 101* a positive recommendation and review here in this book, and consequently thought it would peak your interest even more if I quoted one of my favorite pages from it. This quote is from page 37 of the 1990 edition, and is talking about how the history we know is mostly lies and distortions.

"*An example of this can be seen in American history's rendition of its bloody conquest of the West. Every account will tell you that the white man won his war against the heathen and savage Indians. Not only are these reports unabashedly biased (and the lie of separation that fueled the event carefully hidden), but the fact that the allegedly defeated Indian Nation was the true victor is never acknowledged at all."*

You read that sentence correctly. The Indians won their war with the white man. That was a struggle for spiritual ascendancy, not a battle to determine who would subjugate the land. The Indian peoples, who are fifth-dimensional representatives, sacrificed themselves with their very blood to assure that this nation would become the spiritual giant it was meant to be. History books are incapable of recording that fact because their function is to meticulously gather material in support of the fear-filled national egos they represent. Because they serve human self-deception, historic accounts totally ignore the grander and deeper movement of Spirit that is always the only truth behind any fact.

The Planetary Liberation Organization is here to assure that the fifth-dimensional victory the Indian nation actually won is finally demonstrated, and that the third-dimensional wounds sustained in that battle are fully healed. America will then assume its true spiritual identity and discover what Manifest Destiny really means." (Diana Luppi)

This confirms my own belief that the true power of this country comes from its Native American spiritual and cultural wisdom. My first book, *A Medicine Story: The Return to Native American Consciousness*, discusses this at length, in the context of my own personal spiritual journey.

I left Pagosa Springs around 3:30 p.m., and drove over Wolf Creek Pass, stopping once to take photos, since the colors were so beautiful, and just could not decide whether to go through Crestone, where I had camped several times before, or to strike out for a new location, and head toward Walsenberg, and Trinidad. I had sold some of my books to the **Moonlight Bookstore** in Pagosa Springs, and was happy about that, and was tentatively aiming to find a bookstore someone had told me about in Alamosa. So I headed in that direction. However, it was past 6 p.m. by the time I got near Alamosa, and I felt the urgency of finding my next camp spot before it was dark. My truck was also, for some reason, driving as if a tire could go flat, or it was out of alignment. That scared me, as I was driving in fairly rural and isolated areas between small towns, I had about $20.00 cash, and no major credit cards! I couldn't even afford a tow! All I needed was a breakdown, especially as it was going to be dark by 7:30 or 8 p.m.

So, I had a very nervous drive over the vast flatlands of the San Luis Valley. I wished I could have stopped to sell some of my second book, *Beautiful Simplicity: The Journey Into a Hidden Reality*, in that area, as it mentions UFO sightings in the San Luis Valley. So, I drove madly toward one of the few camp sites on the map, just South of La Vita, trying to get there before dark.

It was dark by the time I got to La Vita, and I still had twelve miles to go to find the campground. I knew that I was not too far from Trinidad, and Walsenberg, both areas of UFO activity. Thus, I was also nervous about that, while driving in the dark to the campground. I really don't like night driving, and especially since I can't see the scenery, or know what my final camp site **really** looks like in the light.

I finally got there, and the sign said "Campground 4 Miles" down this dirt road. I could not imagine driving another four miles on a dirt road, as it was nearly 8:30 p.m., dark, and cold. I just decided to pull over in a picnic area, since this time of year everything was deserted, anyway. I did not have any dinner, in my rush to get there. Now it was too dark, cold, and scary for me to want to set up a cook stove and prepare anything warm to eat. So, I ate some snack food, after moving all my stuff into the front seat so I could sleep in the back. In fact, it was so cold and spooky, that I just went right into my truck. I put on some candles, but it wasn't really light enough this time to read or write very well.

Suddenly, a dark truck with one man in front came barreling down the dirt road, heading in the area of the campground, deeper into the park. I heard him coming, instantly blew out my candles, and hid under my sleeping bag! Not that that would do any good, as my truck flap does not even have a latch that works right, let alone a lock. I decided that was an omen to go to sleep. It was hard to go to sleep at 8:30 p.m., but it was too cold to do much else.

Just before the man drove by, I had gotten the thought in my mind, almost like a message, "You will have the scare of your life tonight."" Then the man drove by, so I was not going to take too many chances. I thought it could mean I would have some kind of an "encounter" with a UFO or aliens. I was not ready for that. So, I slept, to put it all out of my mind.

The next strange thing was that it seemed I heard a motor

going during the night, like possibly a motor boat on a lake. There may be a lake in the area, but I wasn't sure. And, then why would they be out at night, when it was so cold out this time of year?

Then, at 2:30 a.m., suddenly I was awakened by something brushing the front or top of the truck! I could not tell what it was, and I could not see anything out of the side window I was facing. I did not want to go looking for anything, so I laid still. For a couple of minutes, the sound continued. I thought it was perhaps a squirrel or racoon walking on top of my truck, or on the front part. I did not hear any footsteps. It was spooky!

Then the sound abruptly stopped, and I calmed down, and went to sleep.

In the morning, near the stream that I was camping close to, I did see hoof prints, and later that morning, saw cows! Once again, people were using public land to graze their cows. I even had cows on my second vision quest. They were turning up every where on this trip, too. Still, I don't know if what brushed my truck was a cow, because it seemed to be on top of the truck. So, I'll never know.

After this adventure, I said to myself, "That's enough of dark, spooky, cold nights sleeping in my not-so-comfortable truck! I'm going home today."

It was another beautiful day for a drive. I stopped in La Vita for a $3.00 shower, had breakfast in Walsenberg, and continued up I-25 to a turn off by Colorado City. Past there I drove through a National Forest, and stopped for a lunch snack in a lovely park along the way. That drive was also memorable. I took more photos.

After this, I just had to keep driving, and it was becoming tedious. I had to go back the way I had come, up 285, and as I was nearing Breckenridge, it started getting cloudy. The trees nearer home were more bare, and less colorful, and it was getting colder. I guess it was time for this vacation to end.

Coming down Wolf Creek Pass toward Pagosa Springs, CO

CHAPTER FOUR:

THE ESTABLISHMENT OF A NEW VIEW OF HUMAN HISTORY

This is a time in human history when many individuals living in technological societies are coming to an awareness that the "Establishment" version of history is a white-wash job or cover-up. The UFO cover-up is the lead in for people to begin to investigate and to realize how extensive the cover-up process really is. It takes a determined mind to delve into the mysteries that are being hidden from the mass population. It also takes a courageous heart to even begin to face the extent to which the public are lied to, used, and manipulated by their Government(s) and the media. Of course, just my saying this does not prove it is so, but in my mind, the reading I have done in recent months, has convinced me that we are being fed what some very powerful people, with agendas of their own, want us to know. The hidden information is not just about what is going on today in contemporary society, but extends back into ancient history. In fact, what we don't know, or are not being told, about ancient history may be the **reason** for the UFO cover up that exists today.

The "cover up of ancient history" concept had already dawned on me when I began to read *NEXUS New Times* magazine, which originates out of Australia. From subscribing to this magazine (address in earlier chapter), I found ads for the "Adventures Unlimited" book and video catalog, put out by author and adventure-er David Hatcher Childress. As I could afford to, I ordered books, cassettes, and videos from this source. (**ADVENTURES UNLIMITED,** P.O. Box 74, Kempton, Illinois, 60946, USA. And now www.AdventuresUnlimitedPress.com.)

The first book by Childress that I ordered *was Lost Cities of North and Central America.* In this book, he recounts an astonishing story that points to the cover-up of an archaeological find of Egyptian relics in the Grand Canyon!

Through some friends, Childress discovered the micro film of a front page article from the April 5, 1909, *Phoenix Gazette*, which had headlines saying, "Explorations in Grand Canyon," "Mysteries of Immense Rich Caverns Being Brought to Light." "JORDAN IS ENTHUSED," and "Remarkable Finds Indicate Ancient People Migrated from Orient."

Since the article itself is so old, there is no copyright on it, so I am quoting it here, from Childress' book, page 317--320. (I'm sure David won't mind, as I am thus giving him more free advertising.)

"The latest news of the progress of the explorations of what is now regarded by scientists as not only the oldest archaeological discovery in the United States, but one of the most valuable in the world, which was mentioned some time ago in the GAZETTE, was brought to the city by G.E. Kinkaid, the explorer who found this great underground citadel of the Grand Canyon during a trip from Green River, Wyoming, down the Colorado River, in a wooden boat, to Yuma, several months ago.

"According to the story related to the GAZETTE, the archaeologists of the Smithsonian Institute, which is financing the explorations, have made discoveries which almost conclusively prove that the race which inhabited this mysterious cavern, hewn in solid rock by human hands, was of oriental origin, possibly from Egypt, tracing back to Rameses. If their theories are born out by the translation of the tablets engraved with hieroglyphics, the mystery of the pre-historic peoples of North America, their ancient arts, who they were and whence they came, will be solved. Egypt and the Nile, and Arizona and the Colorado will be linked by a historical chain running back to ages which stagger the wildest fancy of the fictionist.

"Under the direction of Professor S.A. Jordan, the Smithsonian is now pursuing the most thorough explorations, which will be continued until the last link in the chain is forged. Nearly a mile underground, about 1480 feet below the surface, the long main passage has been delved into, to find another mammoth chamber from which radiates scores of passageways, like the spokes of a wheel. Several hundred rooms have been discovered, reached by passageways running from the main passage, one of them having been explored for 854 feet and another 634 feet. The recent finds include articles which have never been known as native to this country, and doubtless they had their origin in the orient. War weapons, copper instruments, sharp-edged and hard as steel, indicate the high state of civilization reached by these strange people. So interested have the scientists become that preparations are being made to equip the camp for extensive studies, and the force will be increased to thirty or forty persons.

"Before going further into the cavern, better facilities for lighting will have to be installed, for the darkness is dense and quite impenetrable for the average flashlight. In order to avoid being lost, wires are being strung from the entrance to all passageways leading directly to large chambers. How far this cavern extends no one can guess, but it is now the belief of many that what has already been explored is merely the "barracks," to use an American term, for the soldiers, and that far into the underworld will be found the main communal dwellings of the families. The perfect ventilation of the cavern, the steady drought that blows through, indicates that it has another outlet to the surface.

"Kinkaid was the first white man born in Idaho and has been an explorer and hunter all his life, thirty years having been in the service of the Smithsonian. Even briefly recounted, his history sounds fabulous, almost grotesque:

"First I would impress that the cavern is nearly inaccessible. The entrance is 1,486 feet down the sheer canyon wall. It is located on government land and no visitor will be allowed there under penalty of trespass. The scientists wish to work unmolested, without fear of the archaeological discoveries being disturbed by curio or relic hunters. A trip there would be fruitless, and the visitor would be sent on his way. The story of how I found the cavern has been related, but in a paragraph: I was journeying down the Colorado River in a boat, alone, looking for mineral. Some forty-two miles up the river from the El Tovar Crystal Canyon, I saw on the east wall stains in the sedimentary formation about 2,000 feet above the river bed. There was no trail to this point, but I finally reached it with great difficulty. Above the shelf which hid it from the view of the river, was the mouth of the cave. There are steps leading from this entrance some thirty yards to what was, at the time the cavern was inhabited, the level of the river. When I saw the chisel marks on the wall inside the entrance, I became interested, securing my gun and went in. During that trip I went back several hundred feet along the main passage, till I came to the crypt in which I discovered the mummies. One of these I stood up and photographed by flashlight. I gathered a number of relics, which I carried down the Colorado to Yuma, from whence I shipped them to Washington with details of the discovery. Following this the explorations were undertaken.

"The main passageway is about 12 feet wide, narrowing to nine feet toward the farther end. About 57 feet from the entrance, the first side-passages branch off to the right and left, along which, on both sides, are a number of rooms about the size of ordinary living rooms of today, though some are 30 by 40 feet square. These are entered by oval-shaped doors and are ventilated by round air spaces through the walls into the passages. The walls are about three feet six inches in thickness. The passages are chiseled or hewn as straight as could be laid out by an engineer. The ceilings of many of the rooms converge to a center. The side-passages near the entrance run at a sharp angle from the main hall, but toward the rear they gradually reach a right angle in direction.

"Over a hundred feet from the entrance is the cross-hall, several hundred feet long, in which are found the idol, or image, of the people's god, sitting cross-legged, with a lotus flower or lily in each hand. The cast of the face is oriental, and the carving shows a skillful hand, and the entire is remarkably well preserved, as is everything in this cavern. The idol most resembled Buddha, though the scientists are not certain as to what religious worship it represents. Taking into consideration everything found thus far, it is possible that this worship most resembles the ancient people of Thibet. (article's spelling.) Surrounding this idol are smaller images, some very beautiful in form; others crooked-necked and distorted shapes, symbolical, probably, of good and evil. There are two large cactus with protruding arms, one on each side of the dais on which the god squats. All this is carved out of hard rock resembling marble. In the opposite corner of this cross-hall were found tools of all descriptions, made of copper. These people undoubtedly knew the lost art of hardening this metal, which has been sought by chemists for centuries without result. On a bench running around the workroom was some charcoal and other material probably used in the process. There is also slag and stuff similar to matte, showing that these ancients smelted ores, but so far no trace of where or how this was done has been discovered, nor the origin of the ore.

"Among the other finds are vases or urns and cups of copper

and gold, made very artistic in design. The pottery work includes enameled ware and glazed vessels. Another passageway leads to granaries such as are found in the oriental temples. They contain seeds of various kinds. One very large store-house had not yet been entered, as it is twelve feet high and can be reached only from above. Two copper hooks extend on the edge, which indicates that some sort of ladder was attached. These granaries are rounded, as the materials of which they are constructed, I think, is a very hard cement. A grey metal is also found in this cavern, which puzzles the scientists, for its identity has not been established. It resembles platinum. Strewn promiscuously over the floor everywhere are what people call 'cats eyes,' a yellow stone of no great value. Each one is engraved with the head of the Malay type.

"On all the urns, or walls over doorways, and tablets of stone which were found by the image are the mysterious hieroglyphics, the key to which the Smithsonian Institute hopes yet to discover. The engraving on the tablets probably has something to do with the religion of the people. Similar hieroglyphics have been found in southern Arizona. Among the pictorial writings, only two animals are found. One is of prehistoric type.

"The tomb or crypt in which the mummies were found is one of the largest of the chambers, the walls slanting back at an angle of about 35 degrees. One of these are tiers of mummies, each one occupying a separate hewn shelf. At the head of each is a small bench, on which is found copper cups and pieces of broken swords. Some of the mummies are covered with clay, and all are wrapped in a bark fabric. The urns or cups on the lower tiers are crude, while as the higher shelves are reached, the urns are finer in design, showing a later stage of civilization. It is worthy of note that all the mummies examined so far have proved to be male, no children or females being buried here. This leads to the belief that this exterior section was the warriors' barracks.

"Among the discoveries no bones of animals have been found, no skins, no bedding. Many of the rooms are bare but for water vessels. One room, about 40 by 700 feet, was probably the main dining hall, for cooking utensils are found here. What these people lived on is a problem, though it is presumed that they came south in the winter and farmed in the valleys, going back north in the summer. Upwards of 50,000 people could have lived in the caverns comfortably. One theory is that the present Indian tribes found in Arizona are descendants of the serfs or slaves of the people which inhabited the caves. Undoubtedly a good many thousands of years before the Christian era a people lived here which reached a high stage of civilization. The chronology of human history is full of gaps. Professor Jordan is much enthused over the discoveries and believes that the find will prove of incalculable value in archaeological work.

"One thing I have not spoken of, may be of interest. There is one chamber the passageway to which is not ventilated, and when we approached it a deadly, snaky smell struck us. Our lights would not penetrate the gloom, and until stronger ones are available we will not know what the chamber contains. Some say snakes, but others boo-hoo this idea and think it may contain a deadly gas or chemicals used by he ancients. No sounds are heard, but it smells snaky just the same. The whole underground installation gives one of shaky nerves the creeps. The gloom is like a weight on one's shoulders, and our flashlights and candles only make the darkness blacker. Imagination can revel in conjectures and ungodly daydreams back through the ages that have elapsed till the mind reels dizzily in space.

"In connection with this story, it is notable that among the Hopi Indians the tradition is told that their ancestors once lived in an underworld in the Grand Canyon till dissension arose between the good and the bad, the people of one heart and the people of two hearts. Machette, who was their chief, counseled them to leave the underworld, but there was no way out. The chief then caused a tree to grow up and pierce the roof of the underworld, and then the people of one heart climbed out. They tarried by Paisisvai (Red River), which is the Colorado, and grew grain and corn. They sent out a message to the Temple of the Sun, asking the blessing of peace, good will and rain for the people of one heart. The messenger never returned, but today at the Hopi villages at sundown can be seen the old men of the tribe out on the housetops gazing toward the sun, looking for the messenger. When he returns their lands and ancient dwelling place will be restored to them. That is the tradition. Among the engravings of animals in the cave is seen the image of a heart over the spot where it is located. The legend was learned by W.E. Rollins, the artist, during a year spent with the Hopi Indians. There are two theories of the origin of the Egyptians. One is that they came from Asia; another that the racial cradle was in the upper Nile region. Heeren, an Egyptologist, believed in the Indian origin of the Egyptians. The discoveries in the Grand Canyon may throw further light on human evolution and prehistoric ages."

The article mentioned how large the caves were--that 50,000 individuals could have lived there! One may wonder, why did so many people, a city in fact, need to live underground? Was there an enemy, or some kind of environmental danger, that prompted them to live underground--where they would have to create some kind of artificial lighting and heating to function? What was the purpose of an underground civilization? They would also have to haul their food into the caverns from where ever it was cultivated. (However, if the river was up near the entrance to the cavern, hauling grain and supplies by barge would actually have been easier than using land transportation.)

The Hopi Indians have legends and creation stories which state that they emerged from an underground world to the above world from a point of emergence, called a Sipapu, which their legends locate in the Grand Canyon! The Hopi settlements and reservation are to this day very close to the Grand Canyon in Arizona.

David Hatcher Childress further investigated this mystery, and discovered the strange coincidence that "much of the area on the north side of the canyon had Egyptian names." (Pg. 324, same source.) There were some Hindu names as well. David inquired with a State archeologist, and found out that there **is** an area of Grand Canyon National Park that is sequestered off, and even rangers are not allowed in this region. When he asked why, he was told, "because of dangerous caves." (Pg. 324.) The early explorers just liked Egyptian and Hindu names, he was also informed. A phone call to the Smithsonian Institution brought only a denial that any Egyptian artifacts had ever been discovered in North or South America. (Pg. 322.)

Just after I had decided to include this information in this chapter, other references to Egyptian relics started turning up in my daily life. A *FATE Magazine* article in the January 1995 issue was titled, "Egyptian Mummies in Mexico." (By John H. Tierney) In 1944, Waldemar Julsrud discovered 33,500 clay and stone artifacts in burial sites in cliffs near Acambaro, Mexico. They were dated by experts as possibly being from as far back as 4500 B.C. There was an attempt to prove this discovery as a hoax, yet no proof ever revealed itself. Even so, the author states, "My Freedom of Information Act investigations revealed that taxpayer-financed Smithsonian Institution officials, who had professional conflicts of interest, dispensed false information on the controversy, and funneled official files to the chief hoax claim instigator." (*FATE Magazine*, Jan. '95, pg. 43.)

This article shows a photo of the German discoverer, Waldemar Julsrud's hand holding a very Egyptian-looking artifact. Many of the artifacts had strong Egyptian style parallels. In fact, many books, not well publicized, have been written claiming that the Egyptians had traveled to many parts of the globe, including what is now Mexico. These artifacts, and the site where they were found, have been suppressed and ridiculed, rather than studied, by the professional archeologists. As David Hatcher Childress has suggested, perhaps we have a "Smithsonian-Gate" type of cover-up of ancient relics and information which would destroy our current theories of the history of humanity.

And yet, I know that I am scratching only the tip of the iceberg with this information.

Next, I picked up a copy of *THE LOST REALMS,* by Zechariah Sitchin. This book, too, found some fascinating parallels between archeological sites in South America and in Egypt.

After the Egyptian theme was showing itself, some more synchronicities occurred. While working at Journey Books one Saturday, I saw the latest book by Mary Summer Rain, who has written many books on Native American Spirituality and related topics. The new book was her first novel, and was titled, *The Seventh Mesa*. The book jacket told of an archeological discovery in New Mexico, in which ancient Egyptian relics, and relics from other cultures, were found. While the book is supposed to be fiction, knowing Mary Summer Rain's work, no doubt she was creating a vehicle to reveal information she has speculated on and researched. And, was this discovery, in her story, found inside a pyramid buried in New Mexico? I will have to read the book yet, to get the details, but even from the book jacket, I could tell there was a significant synchronicity here.

The next Egyptian thing that happened was when a man who had bought and read my book, *Beautiful Simplicity: The Journey*

Into A Hidden Reality, wrote to me and told me to immediately go see the movie just then out, **STARGATE**. He said that it showed many ideas that could relate to some of the information I had shared in my book. And, there are some more coincidences here, too.

In the movie **STARGATE**, an Egyptian relic is discovered in 1928, in the Giza area. A little girl, the daughter of an archeologist, was there. Later in the movie she is an old woman, still trying to solve the mystery of what the relic means. The fact that it was discovered quite a few years ago, and had been kept secret, and later hidden in a secret military base in Colorado, seems to parallel the timing and situation of the artifacts discovered in the Grand Canyon. There are secret, underground military bases in Colorado.

Then, a talented, young, Egyptologist, who has unconventional ideas that draw ridicule to himself, is hired by the old woman, who had been there when the artifact had been discovered. He is taken to the secret Colorado military base, where he finds a way to translate the diagrams on the object, which is like a large hoop, and becomes like a round doorway, if stood on its end. The translation enables them to activate it, and it turns out to be a doorway through which travelers can go to another galaxy.

Once in the other galaxy, on an unknown planet, a pyramid, just like the one in Egypt, is discovered. What ever civilization had existed there had degenerated, as the dark-skinned natives of the area are now illiterate, and do not even understand their own hieroglyphics, which still exist on some ancient walls.

Later, it turns out that some evil alien creature has them enslaved, and they work doing mining for the aliens. The aliens have a ship that lands on top of the pyramid. There is a hint, at the end of the show, that the evil aliens just use human genetics to create bodies for them to inhabit. In one split second scene, the real culprit looks very much like the so called "grey" alien, of the type seen in the movie, **FIRE IN THE SKY.** It looks like the same exact model or costume was used.

That the "Stargate" was a military secret, and that we felt it should be destroyed, to keep out the evil invaders from our planet, is suggested by the story. That there is a connection between the Egyptian pyramids, and a method of traveling to other places or dimensions in the Universe, is suggested by having the parallel

Egyptian-like relics in both worlds which the "Stargate" connects together. That an alien race may have built the pyramids, as well as left behind a "Stargate," is the kind of information, which if true, would no doubt be classified military information.

And, we have a parallel, here, too. Richard Hoagland has several videos out now detailing the sphinx-like face, and pyramidal structures found on Mars from the photo mission done sometime in the '70's. The connection between the artifacts on Mars, as seen in these photos, and Egyptian archeological remains, is fairly implicit.

And, what about the "mysterious" disappearance of the most recent camera probe mission to Mars in September of 1993? The military may have discovered something which they did not want revealed to the population at large. Therefore, the information was covered-up, and the press was not told the truth. This is what many people think happened, including Hoagland. (Jan. 13, 1995.)

As of January 22, 1995, I have now viewed Hoagland's newest, third, video, **THE MOON/MARS CONNECTION.** In this video he finds possible evidence for "ancient alien artifacts" on the Moon, in early photographs taken by NASA. He had to go back in time to when there had been more openness to the public regarding space missions, and many photos taken. However, years later, we now have the computer technology to enhance these older photos to make out strange things on the photos that should not be there, by our current theories of the barren surface of the Moon. He finds what he believes may be the remains of a huge, glass, geodesic dome on the Moon! This, plus a few other strange anomalies in some photos, convince him that the Moon has some kind of ruins on it, which he presumes are "alien artifacts."

He also presents a theory about why NASA, and the Government, may have begun to hide certain information, and initiated a "cover-up" of certain photos. Essentially, there were scientists, sociologists, and others, who were afraid of the "dis-integration" of our culture if these new facts were to be revealed. I admit that Hoagland's presentation seemed cryptic and technical, using a lot of geometry to make a point, and yet there **were** strange markings on certain photos. And there did seem to be some evidence for a cover-up.

I might take the possible discovery of the ruins of a huge glass geodesic dome one step further, with my own speculation. Perhaps these ruins are neither "ancient" nor "alien"? In the book, *Alternative Three*, by Leslie Watkins and David Ambrose, based on a British television program called **Science Report**, that was aired in 1975, there is mention of a secret joint space program between the Super-Powers which has established a base on the Moon, and was in the process of colonizing the planet Mars. Supposedly, the Apollo program was a "smoke-screen" for this more secret space program. Scientists were disappearing in a "brain drain" situation that left some mysteries about where they might have gone. Family members were often mystified, as their relative had disappeared without a trace, or had left phoney forwarding addresses. This was the initial basis for investigation in the film.

However, it turned out that this episode of **Science Report** was staged, and said to be intended as an April Fool's episode. Yet, it was not aired on April 1st. After much research, Jim Keith, who has written a follow-up book called, *Casebook On Alternative Three*, feels that the film version, and the book *Alternative Three* were not factually true, yet were based on speculations and clues that could very well have been true. It was more like, the real names had been changed, and therefore the episode appeared to have been fictional. However, there still appeared to be a cover-up of this program, and it is still very difficult to find, and is never aired on television, especially in the United States. The ideas revealed in it are still probably too close to the truth.

I probably should re-read Jim Keith's analysis of the Alternative Three concept, and do a more in-depth comment on it later on in this book. But, for now, the thing that struck me about Hoagland's third video was that the ruins he may have found on the Moon in his selected photos, could very well be the ruins of a Moon base mentioned in the book *Alternative Three*! This Moon base, called the Archimedes Base in the book, (pg. 167 of 1978 printing by Sphere Books, Ltd., London) was supposedly destroyed by some people who had revolted, and found a way to destroy its protective dome! The reason for the revolt was that innocent people were being kidnaped from the Earth and then subjected to drugs and surgery in order to make them compliant workers for the new Mars colony. Another group of Elite scientists, engineers, psychologists, artists, and other talented representatives of humanity, were being prepped to become the new seeds of the human race on the Mars colony. This was part of the scheme in "Alternative Three" to preserve humanity, in the case of the future un-habitability of the Earth.

In either case, ancient or recent ruins on the Moon, both situations would blow the minds of most people if we heard this on the evening News. Thus, Hoagland is doing the right thing to speculate on the ruins being ancient. That at least puts some distance between us and "them." Yet, those who left ruins of a base on the Moon could even have been humans from an advanced civilization that may have once existed on the Earth. They may not have been "aliens." Humanity could have degenerated after some cosmic war or cataclysm. We could even have destroyed ourselves. Theories of Atlantis speculate on this.

I also think of Shirley MacLaine's channeled reading that stated her boyfriend had been an Ambassador in Atlantis who dealt with Extra-terrestrials. We may have had the same space technology as the "Extra-terrestrials" he had to negotiate with. The true history of our Earth is probably at least partly known by the Secret Governmental agencies, and they do not feel we are ready to know it. Plus, they can maintain more control over us if we are kept ignorant.

I am writing most of this chapter in January 1995. On January 15th, I took a Swedish friend of mine to the Indian Market that we have periodically here in Denver. Native American writer Dr. Chuck Ross was there with a table to sell his books. I waited for a moment when he was free, and went up and shook his hand, saying, "Remember me? You ordered a copy of my book, and now I already have a second book out, on UFOs and Government Secrets."

Just at that moment, as if by some kind of Divine timing or synchronicity, a man came up from my right who had also been waiting to talk to Chuck Ross. He instead began to speak to me. I don't even remember exactly what he said, but he responded something about some videos on UFOs that he was selling. He said that he had worked for the Government at a Level 6 underground installation, where the genetic experiments had been conducted on humans. He had been in charge of bringing them in. Some huge number of people per year were being kidnaped for these illegal experiments. (This is the other side of what happens to those who are "missing persons" -- some go to Mars, and some stay here.)

This man was named Professor Andy Anderson, and he lived in the Denver area. I was amazed at the synchronicity. In an amazed, but almost rehearsed way, I said spontaneously to him, right in front of Chuck Ross, "There--more clues for my third book, already," in response to Andy's opening comments. Then we retired to the middle of the pathway, so others could meet Chuck. Here, Andy and I talked for maybe 15 minutes or more, and he gave me a flyer on his videos, and told me much about them. He was anxious about not having gotten a booth to sell his videos and a gadget that was supposed to increase gas mileage. (I'm not sure if they would have let him, anyway, as the Indian Market is mostly for Indian things, and arts and crafts, and is very competitive for space.) He had a friend who was supposed to have shown up with more flyers for him to pass out, and that didn't materialize, either. But, I was lucky enough to get one listing his videos, so I could order later on, when I had the cash flow for it, and have his address.

I had already heard of many of the things Andy told me, but I was impressed by his certainty, and the coherence of his story. Any question I could ask, he had a ready answer, with no hesitation. He seemed nervous, and full of energy, and open to sharing what he knew.

Andy said he had worked for the Government at a Level Six Underground Base where he had to bring the captured people to the genetic experiment labs. Thousands of people have been abducted and used in experiments. They have vats of genetically created beings and creatures. He said workers were told to stay away from the people and creatures who were there, that they were "insane." In reality they had been given drugs, and were being used to test psychiatric and mind-control medications.

He eventually quit, out of horror, after having discovered the truth. Since then he has been shot at, and lost his Government pension. So, now his only choice was to make and sell the videos, and to reveal this information. In a later conversation this same day, I ran into Andy again, and he said that one of his jobs had been to do filming and make videos for the Government. He had a security clearance, and was called on to go film whenever there was some unusual UFO activity.

I mentioned the movie **STARGATE**. He said that it had a lot of truth in it. Andy said that he had also been in Egypt. He knew a lot about that, too. "The US has put a "capstone" inside the Great Pyramid," he said. "There is a "capstone" inside the Pyramid with a date on it, 1800-something, and "Made in New York," inscribed inside it." He talked so fast it was hard to get the details. "The pyramid had been a launch pad," he said! "There is a whole city under the Sphinx!" "There are cities underground all over the place. Mt. Shasta is one. There is a tunnel system all over, too, from Egypt to here."

The Egypt connection aroused my curiosity. I told him at one point about the information from David Hatcher Childress about the tunnel system with Egyptian and Oriental artifacts found in the Grand Canyon, in an area that even Rangers were not allowed into, and that this information came from a newspaper article from Phoenix in 1906! It had been obviously covered-up since then. It was a tunnel system that could house 50,000 people!

Andy had not heard of this article, but he whipped out a folder he was carrying, which had a map of the underground tunnel system, and it had a circle right where the Grand Canyon would be, which meant, "Possible entrance to the Inner Earth." That seemed to confirm my suspicions. Andy also believed in the Hollow Earth theory. His map showed a lot of underground tunnels in Colorado and New Mexico. I had seen one like it at the UFO Society once.

Andy Anderson also had a video about Nazi UFOs. He said, "We've had anti-gravity technology since Hitler in WWII, who had been getting technology from the aliens, such as laser technology. We had to beat them to get the alien technology."

He's seen and filmed a Pleiadian ship up close, and said it was in one of his videos. "The Pleiadian's can time travel, and I have, too. It takes about three hours for one of their ships to enter the Solar System. The ships use energy from the Ethers."

"The white races derive from the Pleiadians; the Indians all derive from the Stars, various places, and the Black people are from the Earth, genetically created from apes." (Blacks might not like that idea, but that's what Andy told me.) He didn't mention the Oriental races. Maybe they are of the same origins as the Native Americans.

One kind of alien was interbreeding with us, (with the Government's knowledge and permission, I've heard) to save their race, which was dying out. They have given many people implants so they can trace them. "They follow the ancestors," he said. By this he meant that they stick with the same family line. (It's probably just easier, and they know what they are getting, genetically.)

I told Andy that I met a psychiatric patient who said he had been on the ships, and had done "hit jobs" for the Government. The most convincing part of his story, I told Andy, was when he actually read my mind, as I had only **thought** my questions about whether he had seen any Secret Technology or aliens during the course of his Government work! (The whole story of this is in my second book, *Beautiful Simplicity: The Journey Into a Hidden Reality*.) I asked about an E-40 level in Government, and Andy said it wasn't really that high, that his was higher, and he used an earlier letter in the alphabet, maybe C-something, for his. I was curious because the patient had bragged that he had an E-40 Security Clearance. But Andy responded to the psychic idea by stating, "The aliens can make you psychic."

Andy also said that some people had found an entire buried city under Lake Erie. (I had heard, also under Lake Michigan.) That was being kept secret, too.

And, "Denver is the hub for illegal drugs being brought into the country. When Bush was President, they were trying to kill me, for Bush was involved in it. I got shot at several times." (This was the first Pres. Bush.)

I asked him about the Bermuda Triangle. Again, he had an instant answer. "The Bermuda Triangle is a natural phenomena. There are seven such places on the Earth, and there is a time travel quality about them. They are doorways."

"And Atlantis?" I queried.

"Atlantis destroyed itself through mis-use of technology," he immediately responded. He then seemed anxious to get going, to find someone else, or make other connections, and I had things yet to see at the Indian market. So, we parted.

While I made notes on this conversation that very evening, I'm sure I may have forgotten something, and left out a few details, or changed his wording slightly. Still the meaning should be similar to what he meant, unless I totally misunderstood something. Maybe I will talk with him again, and I do plan to order at least one video to check it out. He recommended that I try **THE**

PLEIADIAN CONNECTION first, since I knew about a lot of the other topics already, and did not need to be led slowly to understand it all.

For more information, write to Professor Andy Anderson, Operation Exposure, 2201 Dayton, #208, Aurora, CO. 80010. (303) 340-8695. (I have no idea if he is still there! This book is mainly to be a genuine reprint of the original book, so I leave it in.)

I felt this was a major connection, since he was a real person living near-by, and not a book, even though much of what he said I had read elsewhere. I will have to evaluate this information as this book progresses. Really, I could find out it's all **dis-information** by the end of the book! Who knows what is really going on? Yet, I do feel already that he may very well know some truths that need to be revealed.

As an addition to this information let me add quickly that I was talking with an Indian woman about these things just yesterday, and she said that she felt that there was an underground city under the Black Hills, and that the Black Hills were sacred to the Indian partly because of some connection with the aliens.

I then remembered something about a "wind cave" in the Black Hills area from which the Indians say, they emerged from the Earth thousands of years ago! It is part of their Creation Story; I heard it somewhere in my researches on Indians. So, that would be another entrance to the Inner Earth, which the Government has control over, which may be one reason why the Government does not want to give the land back to the Indians!

There is also a similar cave in the Colorado Springs area, and supposedly, too, some entrances to underground cities in the Crestone and Walsenberg areas of Colorado. (Jan. 22, 1995)

CHAPTER FIVE:

THE SEDONA CONNECTION

While working at the Denver area New Age Store, **Journey Books,** I began to see copies of a magazine called, *SEDONA: Journal of Emergence*! I would meet people who came to the store who had recently been to Sedona, Arizona, and I would find used paperbacks on the special Vortex Centers said to be located there. Also, a friend had recently mentioned wanting to go there. Thus, Sedona appeared to be sending me its energies of attraction. This was during the late Summer to Fall of 1994.

This Sedona business was not new. In fact, I had been to Sedona once before, but all I did was drive through it, and turn around to get back to Phoenix, as I had an interview the next day. That was back in March of 1990, and I was still trying to find a way to become a Certified Pastoral Counselor with the American Association of Pastoral Counselors. There was an internship in Phoenix that I had applied for, as I was curious about Arizona, and the near-by Sedona area. I had flown down for the interview, and had one day to spend sight-seeing. I rented a car, and drove up toward Sedona, stopping at an Indian ruin along the way, and only arrived at Sedona too late to explore. I drove through it, and part way up Oak Creek Canyon, before turning around to head back to the motel for the night. I didn't want to be too worn out for my interview the next morning.

The interview went well enough, I thought, but by now I was not very happy with the Phoenix area. It was not at all as nice as the mountains in Colorado were, and basically a flat desert, with a few small hills. I could tell it would be hot in the Summer. And the salary for the internship was not all that much. I would basically have to re-locate myself for a year or less, to continue the training I desired, but would be barely surviving, financially. I had, coincidently, just started a new job, and it paid more, and was interesting enough to try and keep for a while. So, a move to Phoenix did not really feel right. It turned out that the agency lost its funding, and the internship was not offered to anyone! But, luckily, they did pay for part of my air fare.

When I thought about it later, my life changed after my brief drive through Sedona. I met Sun Bear that August, got a Sacred Pipe for personal prayer, and had it Awakened by him! Then, a year later, in 1991, I was in a crisis from lack of support toward my ordination by my home church. By November of 1991, I had not only resigned from ministry Candidacy, but also quit the church to take up the Native American Path of the Sacred Pipe. Whew! Going to Sedona takes its toll.

So, by October of 1994, I was ordering books through the *SEDONA Magazine*. I read such titles as, *The Alien Tide*, by Tom Dongo, *The Armstrong Report*, by Virgil Armstrong, and *Sedona UFO Connection*, by Richard Dannelley. All of these books seem to be small press, possibly self-published, just as mine are.

At this same time, my male companion, Don, decided that he would like to see Sedona, as well. He had read that sometimes the Vortexes could provide healing, and he had something he wanted healed. We decided to try and go to Sedona over the Thanksgiving Holiday, as some of the work I was doing would slow down that week. Of course, it would have to be a low budget trip, so we decided to try camping out! Even though it was November, we somehow had the idea that it might not be too cold, as Arizona was farther South. We both had cold weather sleeping bags, and we could sleep in my camper truck, so why not try it?

So, after a precarious drive over Vail Pass, which had had some snow recently, and a flat tire in Utah at 9 p.m. the night before Thanksgiving, we eventually made it to Sedona by dusk on Thanksgiving Day. We found a wonderful place to camp before dark, and it was pleasantly warm that night, and the next day. We had a long hike in Boynton Canyon the first day there, and it was nice and warm that day. The next day, Saturday, we encountered rain, hail, and even some snow! It was getting much colder. We saw most of the major rock formations, for shorter hikes, and then decided to forget camping that night. We drove to Flagstaff to get a motel. The drive to Flagstaff was as icy and precarious as any in the Colorado mountains. In Flagstaff, it was only twelve degrees, and very icy everywhere! I thought Arizona was warm in the Winter! We finally found a motel, and settled in for the night. Now, because of this storm, our vacation had to end, and we would have to drive all the way home the next day, in order to try and outpace the worst of the ice storm, especially going over the mountains. Don was a hero and did most of the driving, and it was very scary near Vail Pass up to Georgetown. There was blowing snow and poor visibility. But we made it, just after 10 p. m.

Strangely, just as Don walked in the door that night, he began to get hives! He was troubled by hives for several days afterward, and needed to get some medicine for it. He thought it was something we ate on the way back, but I thought it could have also been from the stress of that long drive, in the dangerous weather.

While in Sedona, I managed to get The Center in Sedona, on Hwy. 89A to take some of my books on consignment.

Another neat coincidence happened in Sedona. I was looking for bookstores to stop in and sell my books, and Don suddenly saw one and pulled over. I was surprised to find there some brochures by The Secret Information Network. I had already found out about this group, and ordered some books and videos from them, which helped me a lot in gathering information. I had mentioned this resource in my second book, *Beautiful Simplicity*. The older man in the store knew them personally. His name was Al Butler. He also had a book that he had helped put together, and was the printer for, called *Mind and Soul Control*, edited by Kay L. Westermann. We made a trade. I gave him a copy of each of my books, thinking it was worth it to get them out there in such a place as Sedona, and he let me have a copy of this *Mind And Soul Control*, book, plus used copies of *The Bermuda Triangle*, and the sequel, *Without A Trace*, both by Charles Berlitz. Strangely enough, I had never actually read *The Bermuda Triangle*, and had not even heard of *Without A Trace*. I was just starting to formulate what this third book of mine would be about, and was looking at a book on Atlantis, which I almost got. However, when I asked Al which book he would recommend, he pointed to *Without A Trace*, without hesitation. As it turned out, this book had stories about the Bermuda Triangle which alluded to the time distortions that often occur there. It also mentioned The Philadelphia Experiment, and evidence of Atlantis, as well. This book was exactly what I needed to read at this time to fill in some gaps in my research, and to stretch my mind in the direction it needed to go. The *Mind And Soul Control* book also confirmed many of the ideas I had already heard, especially about Government secret mind control operations. Thus, my trip to Sedona did provide the next clues to my research.

Later, in January, I happened to be helping **Journey Books** to move to their new location, when a woman called the store wanting to know who she could talk to about some UFO information she had! She was just visiting from Pennsylvania. By coincidence, I answered the phone, and she happened to tell me that she wanted to report an unusual sighting she had over Memorial Day weekend in 1994. She also said that she knew people who had seen a lot of UFOs in Sedona over this same weekend! Then, I told her that Barbara Marciniak had been in Denver that weekend, although I had missed her channeling sessions, and that I, too, had seen a "weird cloud" sometime in May. I referred her to the Denver UFO Society, also.

CHAPTER SIX:

WOLF MEDICINE: DISCOVERING THE DULCE BASE

As I was thinking about this new chapter, I was praying with my Pipe. My pipe bag has wolf fur on it, and I have, over time, felt some connection with the wolf, as well as with its Trixter version, Coyote, or Little Wolf.

I thought to myself how it was that I seemed to "fit" as having some wolf medicine. The *wolf* in some books, is said to be a trailblazer, to have much *stamina,* to have stealth, and to have the discernment to sniff out who its friends or enemies are. The wolf can be a teacher, writer, or thinker. The wolf has been nearly exterminated, but may yet make a comeback, if humanity helps the wolf to rejuvenate its numbers. The wolf is an individual, yet very loyal to its pack. It can come in all colors--white, grey, brown, black.

I felt in tune with the wolf energies this evening. I felt that I did have the persistence and stamina to pursue and research these complex topics--that I was "sniffing out" the Truth, and developing discernment, and that I would later share the discoveries of my "trail blazing" with the rest of the pack--other humans.

I felt that my intuitive mind, and ability to connect and weave together various information, was a "stealth" or delving quality that I had at my disposal. Not everyone would be able to appreciate my style or conclusions, but the process of my journey was out in the open for all to see. It was the *details* of my journey that my wolf qualities would reveal. I took the journey and marked the trail for others to come after me. The wolf travels far in her journeys, rarely stopping to rest, until the task at hand is concluded. And so it was for me in my relentless, **"X-File"**-like search for the Truth.

Synchronous with this feeling was the arrival in the mail on this same day of more "secret information," provided by a correspondent, who had read my other books, and wanted to help my research. It seemed as if a new "wave" of information and insight was on its way.

This correspondent wrote, "If you obtained any information

which contained disinformation, I think I would be able to tell right off, and I would let you know about it. Since I haven't had to do that so far shows that you have been making the most of your research ability and talent." He also adds, "Most of all, I was glad that you were very careful about preserving my own anonymity for security reasons just like you did in some special instances where you did not refer to a friend by name or address. By doing so, it makes it possible for me to send you information like what I have enclosed for you today. If you use any of this particular material, just say that it was sent to you by an anonymous friend, since I did not write any of it myself. Some of it apparently is classified TOP SECRET so that means that it is probably reliable information to use in your book. Write it up the way you interpret it and use quotation marks wherever necessary in spite of the fact that nobody knows exactly who wrote this material."

He also makes comments like, "You are right about the tunnel systems, too. (...) The 'hollow earth theory' tends to throw readers off the track if they think that our planet is filled with a lot of empty space. Nothing could be further from the truth. The Earth becomes much denser as you go toward the core just like it is with large pieces of hail forming as condensation nuclei in mountain rainstorms or a northern hurricane, if you want to be more picturesque about it. Caverns in the crust of the earth, yes, but nothing holds together below the crust. That's where the hot volcano material comes from. The core of our earth is composed of very dense, compressed, materials like iron, etc."

This same writer adds, "I have also heard about the Dulce Base many times before from many different sources, and it does have several man-made levels inside of it. I can send you some very definitive info on that, too. So Mr. Anderson wasn't kidding about the vats and all. It's not a very pleasant sight and smell, trying to feed Greys much the same way a mother does with an embryo."

He also enclosed an article, with no author or source, about the "Dulce Base." This information looked familiar, so I took a break from my writing to peruse a couple of sources that I have already, which could mention it. The two volume, *Cosmic Patriot Files*, did mention the Dulce Base briefly in the time line at the back of Volume Two. Without an index, or chapter title of that name, I could not quickly ascertain whether or not any in-depth information on the Dulce Base was included in it or not. Then I picked up my copy of *Underground Alien Bases*, by Commander

X (Abelard Productions, Inc.) There was an entire chapter called, "The Dulce Base," yet the article that my correspondent had sent me was not just a copy of it. There was similar information, but not an exact duplication. The article from my correspondent appears to be a summary of information from several sources. So, unless an author with a copyright appears, I guess I can quote the article here.

DULCE BASE

"Perhaps the most widely referenced area that allegedly has a base is 2.5 miles northwest of the town of Dulce, New Mexico. The area is normally referred to by researchers and abductees alike as DULCE. The base lies under a kilometer of rock south of Dulce. (Note: The installation is NOT under Archuleta Mesa, but south of the town of Dulce. The Archuleta Mesa scenario has been proven to be incorrect, and has been leading researchers astray for several years.) Over the past ten years, a lot of data has come in about this facility. That data indicates that there are many other bases with similar function, and that many of the early underground facilities belonging to the Atomic Energy Commission in the 1950's were eventually turned over to projects relating to alien technology. Most apparently were added on to by the forces that occupied them.

"The area around Dulce has had a high number of reported animal mutilations, although the number of mutilations slowed down in the mid-1980's when Lawrence Livermore labs accelerated production of artificial blood. Increasing tension between Government forces and aliens over the last several years might have caused a cessation of artificial blood and other supplies, since both animal and human mutilations (as well as abductions) increased markedly starting in 1987. At present, the frequency of all these events has reached a level reminiscent of 14 years ago when there were tense times between alien and Government personnel.

"Residents of the area report that a certain section of road around Dulce Lake has been the site of several accidents. It is said that drivers report that they see a straight stretch of road where the road apparently curves. Elders of the local tribe report that when they walk by this area they have a similar visual problem. Some of them have fallen in the lake because of this. These same Elders refrain from discussing the animal mutilations. One interesting fact is that the Jicarilla Indian creation myth says that they emerged from underground, and that human beings were created by beings from under the earth.

"Rancher Ed Gomez has a ranch 13 miles west of Dulce. From 1975 until 1983 the Gomez ranch was the scene of most of the cattle mutes in the northern New Mexico and southern Colorado area. The Gomez homestead, which is over 110 years old, has lost over $100,000. in cattle over an eight year period. One of the cases occurred 200 yards behind his home. This rancher also related to researchers that combat ready troops have been spotted in areas that only the Indians have permission to go.

"In 1978 there was an agreement between the Ute Indians in Colorado and the US Government that consisted of the Ute Indian receiving all the territory now occupied along the border between New Mexico and Colorado with the explicit agreement that they would strictly enforce a "no trespassing" policy. It is not possible to cross the Ute reservation without special permission from Tribal headquarters. If caught without permission you are liable for a fine and/or jail and expulsion.

"It has been noted by visitors that Dulce townspeople openly discuss the abduction and implantation.

"A recent field investigation of the area adjacent to the Mesa proved to be difficult. Researchers were confronted by several small hovering spheres that had some sort of electronic emission that made them all violently ill.

"Activity in the area of Dulce, New Mexico began almost at the same time as the Roswell crash - 1947. Studies of the area have confirmed that summer troop movements occurred every year after 1947 for quite some time. The construction of a road into the area was done, and trucks went in and out of the town. Later on, the road was mysteriously blocked by the military and destroyed. Some residents of the area remember signs on the trucks which referred to a company which was later proven to be a cover--no company by that name has ever existed. The initial base was built under the cover of a lumbering project--unfortunately, they never hauled logs, but just large equipment.

"Since the initial activity seemed to occur in the same year as several disk crashes, one might wonder why the military would be carrying on a large construction program of this nature, especially since it occurred some twelve years before the famous RAND

CORPORATION conference on Deep Underground Construction, which occurred in 1959.

"They might have discovered information about the base from analysis of the crashed disks or the reptilian-based entities they found on board. It is possible that this facility was functional earlier than 1947. We do know that the base was made in stages using alien technology. Most everything in the base is controlled magnetically. Even the illumination is magnetically induced.

"What we do know is that the upper levels were built after the lower levels--or onto the lower levels. In other words, a US base was built on top of a pre-existing alien base. This is not the only location where this phenomena has taken place. There are indications that there are situations like this up at the Nevada Test Site and elsewhere. Deep sections of the complex are connected to natural cavern systems.

"We know that the base consists of a many-leveled cylinder with tunnels radiating out in several directions. (i.e. the tunnels in the Grand Canyon, Karen's comment.) There are at least 7 levels. We know (through personnel who have worked there) that the first two or three levels are primarily US Government personnel. We know that senators and astronauts have been brought to the facility and have been shown the first several levels. The facility appears to be a state-of-the-art cryogenetics facility and laboratory.

"After the 2nd level, everyone who enters any level is weighed in the nude, then given a uniform. Visitors are given an off-white uniform. In front of the doors to all sensitive areas are scales, which compare the individual's weight to what it was when they entered the facility. The person's ID card must match the registered weight and code for the area for the person to be able to enter or leave.

"Any discrepancy in weight over three pounds will summon the security people. No one is allowed to carry anything into or out of the areas. Alien symbols are visible all over the facility.

"Research at the 4th level includes research on the human aura, research on telepathy, dreams and hypnosis--all geared toward control and manipulation of other beings. Technology is at a state where they can manipulate the bioplasmic body of human beings and program them through introduction of programmed reactions and images which will induce the desired result--completely control them. They can remove an individual's consciousness, and replace it with another consciousness--either synthetic or alien.

"We know that level 5 has alien quarters there; level 6 is a genetics laboratory using alien technology. It has been referred to as the "nightmare hall" by employees. Many-limbed humans and other creations reminiscent of stories of Atlantean experimentation reside there. It is an attempt to breed slave beings suited for specific tasks. It sounds like an Edgar Rice Burroughs story, yet it exists. One of the employees who worked there was the source for the photographs and materials that were the basis for the drawings and notes referred to as the "Dulce Papers."

"Level 7 is where human children and adults are stored as a source of biological materials. Remember the long ships in the series "V" where all the humans were stored? It's similar to that, but it is more of a production laboratory situation. Humans have been seen stored in clear cylindrical containers over 6 feet in height, suspended in a yellow or amber fluid -- alive and conscious, but unable to scream or say a word --a common observation in this installation, as well as some of the other 26 installations in the Midwestern US.

"In the same area at Level 7 are the production areas for humanoids, the vats containing animal and humanoid body parts. There are pictures and video tape of this facility--there is no doubt that it exists. It is like something out of the Burroughs Martian series--do you suppose Burroughs was a member of a covert group even in his day and knew this activity was going on?"

(Note: your author is not familiar with this Burroughs material.)

"Biogenetic research gained at Los Alamos, New Mexico was funded under the cloak of secrecy. It was combined with alien technology in the genetic sciences to produce expendable biological entities for use as the makers see fit. Cloning of humanoids is part of the natural progress of scientific development along "service to self" lines; that it is going on there is absolutely no doubt. There is also no doubt that political figures in governments have been cloned and reproduced.

"It sounds like the plot from a grade B movie, doesn't it? Well, what is more comfortable for humans to handle in the guise of science fiction is actually based on fact; the main reason that this problem exists to the extent it does is that it is all literally underground and covert.

"Although the base actually has more than 100 exit points, including large air intake ducts near Lindrith, these exits and other areas inside the base are covered by cameras. Everything and everyone is watched and monitored.

"In 1978, a small group of workers at Dulce discovered the true nature behind the facility and a resistance unit formed; a resistance unit is a touchy thing in a base that was estimated in 1978 to have a population which included over 18,000 alien beings. By late 1979, the situation led to a confrontation over the flash-tube weapons that US Government and alien security forces carried. A lot of humans were killed. Members of the NRO's Delta Security team were among the casualties.

"On October 23rd and 24th, 1988, a research party consisting of Dr. John F. Gille, who has a PhD. in Mathematics and Physics from the University of Paris; Gabe Valdez, head of the State Police in Dulce; Edmound Gomez, his brother Manuel, and officer Valdez' two sons Jeff and Matt, visited the Dulce area for an expedition. At 7:51 p.m. they noticed a very bright boomerang shaped object radiating white, blue, and green light. As it approached them, it slowed (obviously under intelligent control), seemed to reverse direction, and stopped. It then sped out of sight.

"On October 24th, the party climbed Mt. Archuleta in order to conduct a search for debris from an alleged crash of an "experimental aircraft" flown by an Air Force General in 1983. This crash was reported in the newspapers for two days as having been a small plane, then it was hushed up. The craft was rumored to be a captured disk flown by Americans. Paul Bennewitz reported at one time that the craft had clipped off a large tree in its descent, had hit another tree, regained altitude, skimmed over Archuleta Peak, and hit a third tree in the valley north of the peak. It was then reported to have hit the ground and flipped over twice before coming to rest." (End of quoted article.)

Along with this article, my correspondent included some MAPS showing areas where the underground bases and tunnel systems were located. I will include some of this material here, as long as it prints clearly enough. (Someone later informed me that the maps and perhaps the article were taken from some other book, so forgive me for not realizing this at the time. It could be from something published by Timothy Green Beckley. Forgive me, Tim, if so. I'm not getting rich with this.)

What is interesting about this information is that it is exactly the same as what Andy Anderson had conveyed to me in our conversation, as well as in the video I had bought from him, **"THE**

PLEIADIAN CONNECTION, and PROF. ANDY ANDERSON, 30 YEARS RESEARCHER IN E.T. AND UFO PHENOMENA." From Andy's talk on the video, it seemed that he may have been one of the employees at Dulce who found out the horrible truth.

I have to confess, however, that I was somewhat disappointed with the first part of his video, **THE PLEIADIAN CONNECTION**, as it was really a copy of a video on the story of Billy Meier, and Billy Meier is currently being discredited strongly by UNDERGROUND VIDEOS, Box 527, Beverly Hills, CA. 90213-0527. Of course, this kind of confusion could easily be part of any disinformation or covert intelligence operation, to keep the searching public uncertain as to what to believe, until many give up trying, in confusion and frustration. In my case, I am not as sure of anything from Billy Meier as I might have thought I was earlier. (But some people still like him.)

There is definitely an organized movement behind the scenes which aims to discredit any popular UFO case. Ed Walters of the Gulf Breeze Sightings also experienced this. His second book, *UFO Abductions In Gulf Breeze*, counters the discreditors of his sightings. The Truth seems very elusive, as Fox Mulder in The X-Files TV show has expressed.

Still, my *stealth*, and *wolf medicine* guides me to try and solve the puzzle.

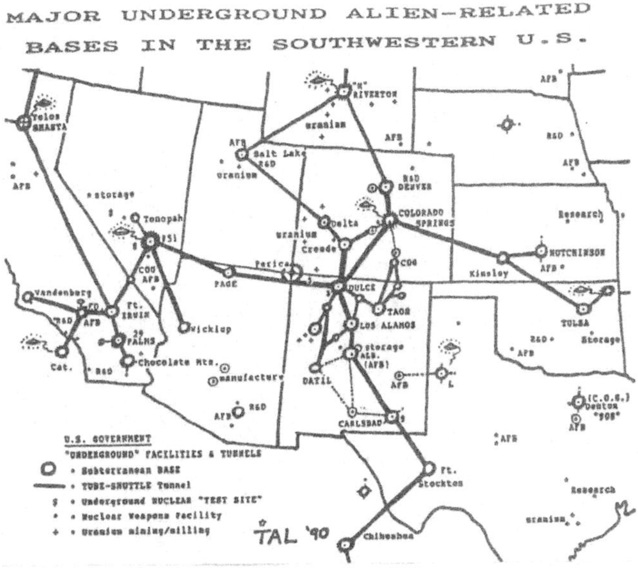

Major Underground Alien-Related Bases in the Southwestern U.S.

Event Characteristics of Four Corners Area

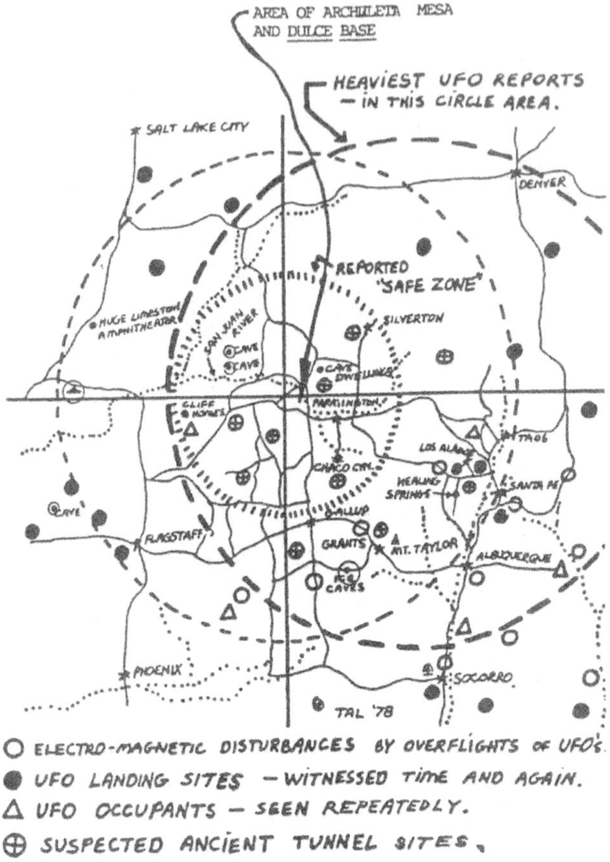

Western Tunnel System

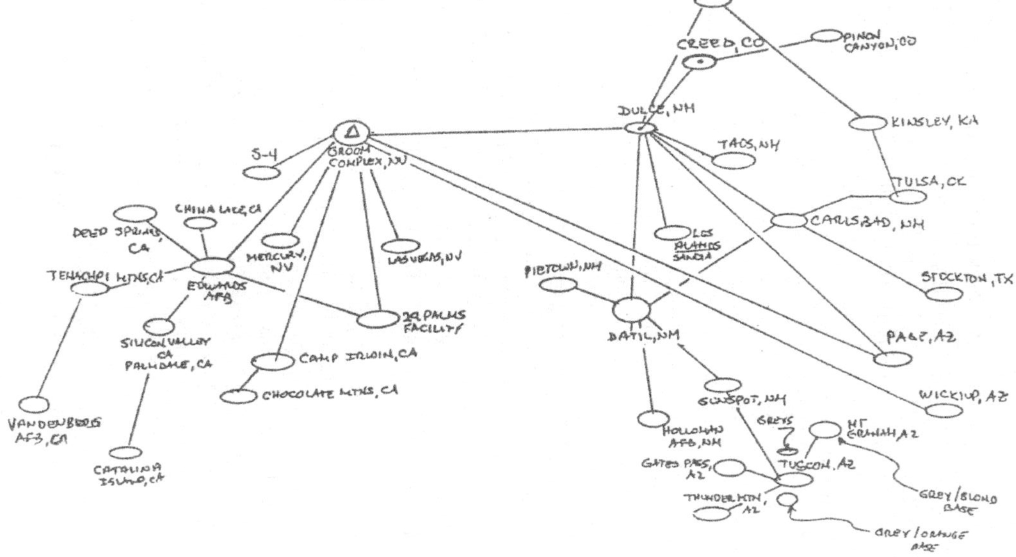

CHAPTER SEVEN:

LOOSE ENDS THAT MAKE IT ALL MORE REAL

Sunday, June 12th, 1994, about 7:10 p.m., I was home alone, watching a nature program on Channel Twelve--the only station our TV can get here on the mountain--when suddenly the TV went off! Then I heard what sounded like a helicopter in the distance. I ran outside in time to see a helicopter go by fairly close over the house, just seconds after I first heard it, and the TV had gone off. I was immediately suspicious that there was some connection between the TV going out, and the sudden appearance of the helicopter.

I tried to get the electricity back on. The house I live in only has solar electricity, so to watch an appliance like the TV the inverter needs to be on. I thought it was still on, but after checking it, the black switch that was on had turned itself off! I touched the inverter and noticed that it was warm, and when I tried to flip the black switch back on--it went off, automatically! Apparently, the inverter had gotten hot and a safety switch had flipped it off. It would not turn on again for about 20 minutes, as my notes on this incident show that I finally got it back on around 7:30 p. m.

When Don got home, I told him that a helicopter flying by had apparently caused the inverter to get hot and turn off. He said that meant that there must have been a large draw on the electricity that over-heated the inverter and shut it off. However, I knew there was nothing unusual on in the house that could have done this.

Not long after this, I read an article in the *SPOTLIGHT Newspaper*, a paper with alternative political information from the mainstream, which indicated that the US Government had "developed electronic technology capable of disabling automobiles and other motor vehicles." ("SPOTLIGHT," Sept. 26, 1994, pg. 5, Liberty Lobby, 300 Independence Avenue, S.E. Washington, D.C. 20003.)

This same article reveals that an electronics firm in Colorado Springs, Colorado, JAYCOR, had already developed a technology intended for police use, called an Auto Arrester. This device could "overload the car's ignition system with an electronic pulse, killing the engine, which could not be restarted."

Don and I have frequently observed several helicopters take a route that goes by our mountain home, especially in the summer. I have often surmised that they were on practice flight maneuvers, possibly from the Air Force Base in Colorado Springs. It is therefore not a surprise that a company in the Springs area was working on a technology such as this. To be able to disable an ignition system, or any other electronic system, using an electronic pulse, is obviously a very powerful weapon to have in battle. One could disable an army of tanks and trucks, cause planes to crash, and prevent bombs from being dropped or shot. Disable the enemy's electrical equipment, and the enemy is a sitting duck.

It seemed very clear to me after reading this article, that there probably was a device such as this in the helicopter that passed over our house that day. Perhaps a soldier in training could not resist trying it out on our strange, solar electric system.

It also came to mind that many times car engines die out when in close proximity to a UFO. Then, after a period of time, they will start up, again. Obviously, a phenomena like this, whether UFO or secret weapon, would have to be kept secret, for security reasons. It was at least 20 minutes before I could get the inverter to go back on again. An enemy would be totally incapacitated for that amount of time.

Other short anecdotes. A friend of Don's suggested we see the movie **AIR AMERICA**. We rented the video. This movie was about secret drug flying and military presence in Laos, before the Vietnam War, or in addition to it. **Air America** was a cover for CIA or other secret drug smuggling done by our Government to do business with certain factions in Laos, and thus help them fund themselves.

Not long after seeing this film, I was visiting my Hopi/Mexican Indian friend, and told him about the movie, **AIR AMERICA**. He then excitedly stood up and took a leather jacket from his office. On the back of the jacket were the words, "Air America"! He had been a pilot on secret missions in the service, himself. That's all he would reveal, but we found that to be an interesting coincidence. He later confessed that the Government wouldn't give him a passport, due to the kind of secret information he has been exposed to. At one point, while in the service, he had been in charge of filming some experiments of some technology that he did not know what it was, and to this day he thinks it may have been some sort of "alien technology."

During July 1994, Don was still working at his pizza waiter job, and one of his co-workers there was a lady who had worked for Naval Intelligence. She said that she knew for a fact that we have UFO tracking systems, and that they are always picking up unidentified blips. She lost her security clearance, and her job, due to her husband's financial problems, and now is divorced, a single mother, surviving by working as a waitress and house cleaner.

These synchronistic bits of information seem to magically appear during the course of our every day lives, as if to confirm the research and reading that I am constantly doing.

Through the circumstances of my life, more clues would appear. One night I was working as an On Call Mental Health Counselor in a psychiatric hospital. I had been reading various articles in the *SPOTLIGHT* newspaper about unusual troop movements and operations within our country. People were also seeing Russian war equipment being moved around the country. A patient at this hospital just happened to mention that housing was now harder to find in Colorado Springs because *20,000 Special Forces troops were moving into Colorado Springs!* Her comment was related to how people were selling their homes now because they could get more money for them, due to this. She did not know that this tidy bit of information could relate to some other over-all pattern of unusual troop shifting across the country. I have even read that Russian Special Forces troops were being trained with ours! (In "Spotlight," again.) Twenty thousand Special Forces troops is a lot. But we don't get information like this on the Network News.

Another day, I was working at **Journey Books,** when a man came in who was involved in Native American Spirituality. He talked about a Master Computer in Brussels, which was going to be the brain behind the New World Order. He also said that another computer in Australia already had artificial intelligence. The final unusual thing he said was that some ruins had been found in the Senoran Desert in Mexico that were a kind of pyramid in which alien bodies had been found! I haven't been able to verify any of this, but the idea of ruins in the Senoran Desert in Mexico with alien bodies in them reminded me of the kind of things David Hatcher Childress writes about in his book *Lost Cities Of North And Central America*, (Adventures Unlimited Press, Stelle,

Illinois.)

Another time at **Journey Books**, a man who I'd talked to previously about his UFO sighting in New Mexico, mentioned that animal mutilations were getting more numerous again in Southern Colorado and Northern New Mexico. (Circa February 1995.) He'd heard that a man who had touched the mouth of one of the mutilated cattle had gotten an acid burn on his hand! And, whereas the earlier cattle mutilations featured cattle that would not decay, these newer ones tended to decompose quickly--apparently due to a highly acidic decomposing agent on them. Supposedly, Black Helicopters were also frequently seen in conjunction with the cattle mutilations.

Another night I worked at a psych hospital in Denver, and spoke with a patient who had been an engineer. He had at one time worked on repairing the launch pads of the Titan rockets. He had had a security clearance. At some point, he began to have "thought problems," was eventually declared schizophrenic, got divorced, and has lived like a hermit couch potato for years now.

In my previous book, *Beautiful Simplicity*, I mentioned a brief TV news clip I saw in January '94, about a "military" Titan rocket that had been launched to "map the moon." That was all we heard. I am surprised that that much was revealed, even. This patient said that there had been 4 Titan rockets sent up. What were their missions? Apparently, it's all Top Secret. We just don't hear much about them.

It makes me wonder if this otherwise highly intelligent man's mental breakdown came about from exposure to something that could have damaged his brain, or traumatized him emotionally? Maybe his security clearance, and connection with the secret space program, gave him more information than he could handle. By his presentation of himself, I was quite surprised that he'd lived like a hermit, with no friends, family, or lovers, and also hadn't worked, for at least eight years. He was articulate and nicely dressed. I get people to tell me the strangest things in psychiatric interviews.

Thus, in the variety of my every day life, I'm always being presented with clues to secret information that may be going on behind the scenes. (3-16-95)

Another time, a patient in a psych hospital told me that when he was in the service, some Tibetan monks came and offered to take a few selected service men to live with them for two years at the Tibetan monastery. He didn't specify which service he was in, or how this arrangement was made, but he was one of those few men chosen for this experience by the monks. This patient learned some meditation techniques, and some spiritual energy rituals. He was allowed to share one of these rituals with a group of patients at the hospital, as an end of the day relaxation activity.

By coincidence, one of my correspondents sent me an article, source unknown, which said: "In 1959, a Soviet mission visited Tibetan monasteries, and were presented with the opportunity to undergo mental training."

This caused me to wonder if the Tibetan monks were training soldiers or civilians of both sides, in their philosophy--perhaps as a step toward world peace, or general public relations. This mental patient was only in for a brief alcohol de-tox treatment, and not because he was mentally deranged. He said that even though it was cold high up in the mountains of Tibet, the monks only wore skimpy outfits, and slept on the ground without any covers!

One wonders for what purpose the Government Service would want to allow some soldiers to undergo this training?

Supposedly, according to the same mysterious article with the unknown source, the soviet people who were undergoing mental training, "...performed an experiment in mental focus at the great monastery of Galden with a lama where a cloudy image came into view. The image was of a creature somewhat like a man, but with features and jointed limbs like those of an insect or crustacean. In front of the creature was a moving reproduction of the solar system, except that there were ten planets instead of nine. There was an additional planet beyond Pluto. The lama refused to say where the images were coming from, but told the scientists that indeed there was a 10th planet beyond Pluto."

The same article says: "The American Apache Indians tell stories of tunnels between their lands and the city of Tiahuanaco, and claim that their ancestors traveled for years by this route. The Indian chiefs also assured that the tunnels were 'carved out by rays that destroy the living rock' and that their creators were 'beings that live near the stars.'"

By coincidence, a copy of *UFO Magazine* which I picked up recently (Vol. 10, No. 2, March/April 1995,) has an article discussing the possible use of "Nuclear Powered Tunnelers On the Moon." Los Alamos inventors have already been given a patent for this device. The tunneler would melt rock as it went along, creating "very strong, vitrified rock walls as it goes," so there is no need for any extra structural support for the "walls and roof of the tunnel." A "glass-walled tunnel lining is automatically formed and melted into place." This article states that NASA has already considered using this tunneler to create underground living areas for researchers on the moon. Some of the glass-like rubbish from the tunneling process could also be used for creating structures on the Moon's surface--an idea which reminds me of Richard Hoagland's recent video in which he believes he has found some glass-like ruins in photos of the Moon's surface. This article is by Richard Sauder.

While I am usually a stickler for at least knowing the present *source* of information, even if it's not a good source, I just wanted to throw out some of the information in this chapter in case someone else may have heard of anything similar.

To conclude this chapter with a neat synchronicity-- on December 8, 1994, I picked a "Medicine Card" related to what I am dealing with in the process of writing this book. The one I happened to get was number 41, "Whale--Record Keeper."

This card fit exactly what this book was talking about--the hidden knowledge of the ancient history of Earth, and how to tap into it.

To briefly quote from the *MEDICINE CARDS* book, "Whale is very much like a swimming library. Whale carries the history of Mother Earth and is said to have been placed here by the Ancients from the Dog Star Sirius." (P.201, *Medicine Cards*, Jamie Sams and David Carson, Bear and Co., Sante Fe, New Mexico, 1988.)

Many people with Whale Medicine, this book says, begin to awaken to their psychic and telepathic gifts when they begin to access the stored records of ancient knowledge. These "records" are actually encoded in their (our) DNA. When a certain frequency of energy begins this awakening process, we begin to "remember" and are led to understand the hidden history of our planet.

Entering Canyonlands National Park, Utah

CHAPTER EIGHT:

NEW REVELATIONS

As I begin to write this chapter in April of 1995, I've been stuck in the house for days because of the nightly Spring snows hitting the Colorado Mountains. It's not that deep, because it keeps melting, but it's often deep enough to force me to park my truck a half a mile away from the house, and the road to town becomes muddy and slick even when the snow is melting. So, if I don't have to go out, I stay home. (This is why I owned my 2 wheel drive Nissan truck; now that I have a 4WD it is easier, in 2010, but not always.)

A couple of months ago, this being trapped inside would have meant that I could not go out to work. I had some cleaning jobs, and was doing on-call mental health shifts in Denver area psych hospitals. Now I can work from home, as I've been hired as a professional psychic and Tarot card reader with The American Association of Professional Psychics, Inc. I just sign in on the telephone line, and take calls from home.

While the calls are sometimes sporadic, I can usually make an average of about $25.00 a day, being on line ten to twelve hours. While this is not riches, it is enough for me to get by while living here in Don's paid for rammed earth home. And I do still work Saturdays at Journey Books. It was through Journey Books that I met the lady named Raven who told me about this psychic line.

While there are times that I don't feel all that psychic, I just have to trust the cards and my intuition. Many times a person will give me feedback, such as, "You described her to a T," when they have asked me to describe someone they are thinking of. To tell the truth, it surprises me when I get accurate information that does not derive directly from the cards. And the cards surprise me as well when they accurately define a current situation a person is in, even when the client gives me no clues. Predicting the future is another matter, as I believe there are times when the future is still in flux, or a person should make decisions based on their gut feelings rather than what a psychic tells them may happen.

This new work has changed my life, as now I have work that I can do anywhere, at any hours I choose, and use both my ability to read cards, as well as my counseling training. If I am not totally sure about something, I can at least fall back on giving good advice, based on having a Masters Degree in Counseling, an MDiv., and a non-traditional PhD. In Pastoral Psychotherapy.

The strange coincidence about all this is that I received my first deck of Tarot cards from my mother, when I was probably in Junior High! That was in the days when there were only a couple different decks to choose from. The Spiritualists I met when I was 16 kept encouraging me to do the cards more. A Spiritualist minister named John Scudder had given me a reading once that told me I would go into a type of work that would not exist for another twenty years, that dealt with interpreting colors and symbols. The computer systems and 900# phone lines that are the base of the psychic lines only came into existence a few years ago. It has been over twenty years now since he gave me that a reading. So, I am back to where I started so long ago, but in a new way. After studying for the Christian ministry, who would have guessed that I would end up being a psychic reader? Yet, my original desire was to become a Spiritualist minister, and they do have to develop and use their psychic abilities to minister to others. I have only become what I have always wanted to be. I also wanted to be a writer, and to tell people of the Mysteries. I'm doing that, too.

At the end of April 1995, there was a "UFO and New Age Expo" in Colorado Springs, set up by Timothy Beckley. I was lined up to be one of the speakers. In exchange for doing a thirty minute talk, I could get in free to the main talks, and could get discounts on some workshops. This was my first such speaking engagement, and I really enjoyed it. But, even more important was that I met some very interesting people there. People such as Vance Davis and Robert Morning Sky.

I had already heard Vance Davis speaking on a tape he had mailed to me, and I had sent him a copy of my book, BEAUTIFUL SIMPLICITY. When I met him at the conference, he gave me a big hug, and a free paperback copy of his new book, UNBROKEN PROMISES. I was happy to have made a new friend. His experiences with his own psychic abilities, as shared in the book, also helped to encourage me, now that my major work was being a psychic.

Vance was one of the "Gulf Breeze Six"–six Army Intelligence Specialists who went AWOL in July of 1990 because of advice and information given to them from some very reliable spirits through a OUIJA Board. One thing the spirits told them was that there was going to be a war in the Mid East, and that they should get out of Europe, and out of the Army, if possible. The spirits told them that this war would last 45 days if Israel stayed out of any direct involvement, otherwise there was danger of nuclear war. When they went AWOL, the six soldiers left copies of the predictions the spirits had told them, which included a message to get this information to the "highest possible command." Presumably, President Bush (the first one) was given the information, as unusual as it was, and he did indeed warn Israel to stay out of the conflict during the Gulf War.

Another prediction was of an explosion in New York City that could be terrorist activity, which came true with the first bombing of the World Trade Center.

The spirits also warned them of a coming "New World Order," which would be evil, and cause the United States Government to fall within ten years. They had never heard of the idea of an evil "New World Order," yet today this idea is so well known that Citizens' Militia are being created in every state by people who feel they will need to defend the Constitution with their very lives, from an internal enemy that will collude with those desiring a "New World Order," one world government.

(Things are just as crazy in 2010.)

The spirits also told them that soldiers would soon be "marked" with micro-chips, and that they should leave the service before they would become "marked." This may be going on already with some soldiers. An article in TIME magazine, May 1, 1995, on Timothy McVeigh, a suspected bomber in the Oklahoma City Federal Building bombing, says that he claims "that the Army had implanted a computer chip in his buttocks." (Pg. 48, TIME.) Perhaps it was the computer chip which contributed to his radical stance, or enabled them to find him so easily. Perhaps he was even "used" to help set up a public suspicion of the Militia, because of his "connection" with them. That is, he may have been manipulated in some way to commit that horrible act. Or he may be another "patsy" like Oswald said he was in the JFK assassination.

The latest issue of THE SPOTLIGHT newspaper, for May 15, 1995, has an article by an "ex-high level FBI official," Ted L. Gunderson, who says that the bomb that hit Oklahoma City was "a very high tech and top secret barometric bomb," and that there were two explosions, a few seconds apart. There is no way that a fertilizer bomb on the street could have done that kind of damage. A seismograph record from the Oklahoma Geological Survey at the University of Oklahoma 15 miles South of Oklahoma City indicated that there were two kinds of explosions, ten seconds apart. This very fact indicates what kind of bomb was used, and clued in the FBI veteran that the "fertilizer bomb" only theory was not true. If this is so, there must be a very big cover up here, and the whole thing could be some kind of operation to blame the militia for this horrible event.

As if to confirm what I read in THE SPOTLIGHT, on May 15th, I attended a Global Sciences meeting in Denver, at which one of the speakers was a man who a had worked for years in nuclear bombs and related information. He also said that there was no way that that horrible explosion in Oklahoma City was caused by a fertilizer bomb. He'd even tried that himself for some mining operations in Colorado, and it would hardly even ignite.

Also, he pointed out, the story kept changing; first it was a bomb in the trunk of a car, then in a van, then a much larger truck of some kind, which would be needed to hold that much fertilizer to create such a blast. Also, he said it was an interesting coincidence that the FBI agents, the Drug and Firearm office (BATF), that had invaded the compound in Waco, TX., and the Cult Awareness Network people all had not shown up for work that morning. Remember from the early new reports, they kept saying there were some FBI agents who were not accounted for? That was sort of a mystery that the reporters never resolved. And their offices were in the back of the building, not near the blast, he said.

This scenario reminds me a lot of the cover-up after the JFK assassination. If you are not sure what I mean, check out the movie, JFK, for even though it may not have the whole story, it has enough to convince one that Oswald did not, and certainly not by himself, kill President Kennedy. I have also heard of the possibility that Waco was a Secret Mind Control experiment, much like the Jim Jones event was, and they had to wipe it out to cover up the whole thing. There are several videos on the Waco incident, and I urge those who seek the truth to view those, too.

Another way to consider the Oklahoma City story given to the public, is that the Government knows it was not just fertilizer, but that the kind of bomb used is so classified that they are making up a cover story to protect the classified information from getting out. I certainly hope that the main stream of our Government did not plan this bombing, just to make the militia look bad, but even if they didn't have anything to do with it, there could be a faction, or a Secret Government, or a combination of inter-Government factions involved. And, someone must have warned the FBI, and other agencies, to tell their people not to go to work that day. If the public is given a snow job on this case, and they believe it without question, we are all in trouble.

(By 2010, Timothy McVeigh was eventually put to death for this crime, yet it still seems like there was more to it. It reminds me of the later controversies over the 911 attach on the World Trade Centers. There are many engineers and architects who still say there was evidence of some sort of chemical explosive in the rubble that made that more like a planned demolition, in part, as well.)

Back to the UFO Expo. The Highlight of the UFO Expo was finally meeting Robert Morning Sky. Robert is a Hopi/Apache Native American man who says that in 1947, while on a vision quest, his grandfather rescued an alien being from a crashed disk, only minutes before soldiers got to the sight and retrieved it. This was about a month after the famous Roswell incident, and occurred in the Four Corners area..

Morning Sky is writing a series of books about the "Star Elder" who had been rescued by a g roup of Indians on Vision Quest, and about the information the Being had communicated to them. The story "Revelations of Star Elder" was also partly published in segments in SEDONA :JOURNAL OF EMERGENCE! Starting in the May 1995 issue.

The reason I say, I "finally" met Robert Morning Sky is that I had been trying to meet him in person ever since that last time I was in Australia to visit my family in November 1992. While there, I had seen flyers with his picture on them, about some workshop he was involved in. Then, I was sure I saw him on the plane when I returned on December 7th 1992. I saw him going through customs at the same time I was, from the same plane.

There was also another man with him then, who I thought looked like a person I had seen photos of, who taught "survival skills," and who also did workshops to teach his knowledge. This person says he has learned a lot from the Australian aborigines, in the ads I saw about him.

I saw them both go through customs together, so they probably knew each other or might have been in the same workshop. (Now, in 2010, I have no idea who this other guy could be.)

While I suspected I was seeing the Robert Morning Sky who I had seen flyers about in Australia, I had no reason to speak to him then.

In 1994, there was a "Self-Reliance" Expo in Denver, which I attended. Robert Morning Sky was expected to attend and give a talk there. I signed up for it, but apparently he never showed up to that one. Or, if he did, it was when I was not there.

Then, I kept seeing his name advertised at some of Beckley's UFO Expos, but I could not afford to go to the ones that were in Pheonix or California., so I wrote Beckley and said that I'd like to try being a speaker at one of his Expos, if I could get in free, and especially if they had one nearer to where I lived.

Thus, my destiny unfolded when Timothy Beckley held an Expo in Colorado Springs in April 1995, and I was invited to be a speaker. Robert Morning Sky was scheduled to attend as well. His topic related to revelations about ancient history that the "Star Elder" had told his grandfather, through a method of communication they had developed.

I knew this would be significant in my own search, as this third book of mine that you are now reading began to unfold as a search for the missing pieces of ancient history. I had no idea when I started this book what it would actually be about, what information I would be led to. All I knew was that a psychic had advised me to write another book.

In Morning Sky's workshop he told a story of the history of our galaxy that was supposed to have been told to his grandfather by the "Star Elder." This was a long and complicated saga, almost like a soap opera. Some of the characters have names that are similar to those used in Zachariah Sitchin's books. (Yet, he never credited Sitchin, but supposely had gotten this same info from his grandfather and the Star Elder.)

The Earth was created, he said, from a remnant of a much larger planet that once existed where the asteroid belt is now. The moon was also a portion of that explosion. There was an atmosphere on Mars, and life there as well, but the same "bad guys" who were angry enough to blow up a planet, also destroyed Mars.

Robert said that there were Reptilians and Canine originated aliens, but they had all evolved to have a humanoid form, with four limbs and a head. From my scribbled notes, I think I can make out that there were two major empires who have been fighting over this solar system: The Syrians from the Sirius system, and those from the Orion system. As I wrote this book, this is all the information I had, but I believe later he wrote some books with more details.

I am coincidently reading the book, THE ORION MYSTERY, by Robert Bauval and Adrian Gilbert, in which a major discovery they have made is the connection between the three major pyramids at Giza, and the three major stars in the Orion system. They have un-covered a strong "stellar" factor in the religious system of ancient Egypt, and that at the time the pyramids were built, they would have been lined up with the stars in the Orion's Belt. There are also four shafts inside the Great Pyramid which would point toward various star systems, including Orion and Sirius. As I read more of this book, it could add another dimension to the history that Robert Morning Sky told us in his workshop.

From that Expo, I learned that Morning Sky was planning on taking some people on a "Vision Quest" experience in the Canyonlands of Utah, where he says you can see strange rock formations which contain signs of some ancient extraterrestrial battles. I have signed up for this, and expect it to be very rugged for me. But, I could not pass by the chance to learn more of the information he possesses. I may have to suffer to get it. This will be in mid June, so my next chapter could be about that experience. This book is written as it happens.

The view from our mountain home.

CHAPTER NINE:

THE "MORNING SKY" CONNECTION

I have returned from the "Vision Quest" in the Canyonlands of Utah, which was guided by Robert Morning Sky, an Apache/Hopi Native American man. I made new friends on this Quest, and have gotten to know Robert better. My Spirit and heart have awakened again toward the Native American Path, but now in a new way.

Now, I realize that I had gotten discouraged by all the Indian politics, especially among Lakotas. I wasn't going to Sweat Lodges as often, since they were so far away from my mountain home, and I wasn't even praying with my Pipe as much. My male companion was not into ceremonies, and I no longer had much chance to socialize in the city with the few friends I still had who followed the Red Road. The Winter and Spring were long, so I could not easily get out for these things. I stayed home and took calls on the psychic line. To go out for a day or an evening would be a loss of income for me, now, and I wasn't really making all that much on the line, just more for less trouble than the other jobs I had been doing previous to it.

Meeting Robert at the UFO Expo where I spoke in Colorado Springs on April 30th seems to have been the fulfillment of part of my original Vision Dream's message. Or at least, there are some strong synchronicities in the Dream, which seem to connect me with Robert. This is how the story goes:

In November of 1992 I visited my real mother Janet Doty in Brisbane, Australia for a month, returning on December 7th. While I was in Brisbane, I had seen flyers about a workshop featuring an Indian man named Robert Morning Sky. A photo of him was on the flyers. While I could not afford to attend this workshop, I remembered his name for future reference. I had no idea really what his work was about. I just pay attention to coincidences.

As it happened, on the flight back to the States, going across the Pacific Ocean, I had seen an Indian man get on the plane. I also saw him getting his luggage to go through customs back in, probably Los Angeles. It looked like the same person. I was pretty sure it was.

Later, I attempted to connect with him at a "Self Reliance Expo" that was in Denver in 1994. He either did not show up, or came at a different time than when he was scheduled, so I missed him then.

By August of 1994, I had been able to print up some copies of my second book, **Beautiful Simplicity,** and a psychic told me to write a third book, even though I wasn't sure I would. As I began to research information about ancient history and archeological anomalies, I began to realize that there was information out there that pointed to Extraterrestrial involvement in Earth's ancient history, as well as a probable "cover-up" of such information by world governments to keep us from knowing the truth of what they already knew. Since much of my writing is about my own process of discovery, I started out this third book without any real outline or plan for how it would unfold. I am amazed as how things seem to fall into place as I proceed. For what I realized in meeting Robert Morning Sky amazed me even more.

During our Utah "Vision Quest" experience, I took some copies of my first two books, **A Medicine Story**, and **Beautiful Simplicity**, in case anyone would want to buy copies. When we did a talking circle one evening, I told about how I came to write my books. **A Medicine Story** was written as my way of sharing and actualizing my Vision Dream, in which a Voice had told me to "Write theology for Indians," as well as that, if I liked history, "there was a history that needed to be written," as well. I confessed that it really was confusing for me, a recent Christian seminary graduate at the time of my Vision, to understand what it meant. I finally concluded that writing my own story of learning about Native American thought and culture was a way that I could "write theology for Indians." Theology was the study of God, and spirituality, so it didn't have to be only Christian, as God was a Universal Spiritual concept.

As I've told this story many times before, I didn't realize until I looked directly at Robert, that he was looking at me in a very understanding and loving way. His eyes were soft and deep. He really appreciated my Vision, and seemed to believe in it as legitimate. He did not criticize me for writing a book about

Indians, either. I suddenly realized, after this, that my little dream, that I was lucky to have remembered so many years ago, may even have been talking about some of his work. He was looking for people to help communicate to the world the stories and information he was now sharing, about the Star Elder, and a hidden history of the galaxy that his grandfather had learned form the Star Elder.

Perhaps the Source of the Voice in my Vision, that I had way back in 1987, some how **knew** that this information needed to come out, and so my spiritual helpers guided me to become curious about Morning Sky just to get me to make this connection. Now, in this book, I am mentioning his work, so that anyone who reads my story, will be led to his. Thus, I am still "writing theology and history" for Indians.

Because of Robert's respect for me, I also gained a renewed interest in my own story and writings, which I had actually been neglecting, and wondered if I should even re-print my Medicine Story book. It's a major expense to re-print it, and I really don't have much money. I have used up some of the income from book sales on other projects, and even used some of my "book money" to pay for the "Vision Quest" with Robert.

I also feel a need to visit my family in Australia again sometime in 1996, though I don't know where I will get the money for that. Robert has connections in Brisbane, Queensland, Australia, too, which is also a weird synchronicity. Perhaps through Robert I could make some new connections in Brisbane the next time I am there, and feel more comfortable in Australia, should I ever need to move or live there. Any friends of his would be good people to know.

The other thing that struck me was that I had dreams about going to Utah way back in 1987, 1988, and in fact, it was telling Tony Shearer in a letter that I had dreamed of Utah that caused him to respond to my letter, and connect with me. Now, I went to Utah with Robert and his group, and had this remarkable feeling of connection with his work.

Also, my Vision Dream was about an Indian man and a white woman getting married, and Robert's companion is a white woman. In the Dream there was a feeling of significance to the fact that they would have children, and even though the children would be of mixed blood, the special DNA that they both carried would live on.

Robert talks about a special strand of DNA in us, that is a gift from some Extraterrestrial race. It gives us our Soul, and our Passion. It makes us more like individuals, instead of like beasts. It gives us the chance to become Masters, like Jesus. He calls this special DNA, the Gift of the Feather.

In 1991 I had a dream that mentioned that I was to preach on the 30th of some month, and the day of the month on which I spoke at the 1995 UFO Expo was April 30th. In the dream, I thought I needed to be prepared with a written-out sermon, but instead wanted to do something more spontaneous. At the Expo, I had a written-out talk, but instead spoke spontaneously, just using the written talk as an outline.

I guess that one of the key things my stories share is how important it is to try to understand and follow the guidance in ones dreams. My dreams have made me into the unusual person who I have become.

While Robert's Utah "Vision Quest" was not a traditional Vision Quest at all, and he did charge a reasonable fee for the experience, I still call it a Vision Quest, because it did have transformative power for my life. It did open me up to my feelings, my Destiny, and, I hope, a new community of friends. It did have the spiritual power of getting me out of a creative slump, and into a more meaningful connection with my Passion. Now I have a new purpose for my life, and new reasons to continue on with the story telling that I have already been doing.

Sometime last year, I had done a reading for myself with **The Medicine Woman Tarot**, and jotted down a message for myself, which I kept near my computer as I wrote. The message was, **"Your stories transform the world! A new Self is emerging."** Just now, I spontaneously picked a card from this deck again, and got, "Exemplar of the Arrows--Master Storyteller." One message from this passage says, "Hear my stories and you will learn my warrior's way. Listen to my life, and you will find the way that is your own to travel." (Page 227, **The Medicine Woman Inner Guidebook**, by Carol Bridges, the self-published edition, by Earth Nation Publishing, Box 929, Nashville, Indiana 47448., 1987.)

Years ago, another Robert, Robert L. Moore, a professor at Chicago Theological Seminary, inspired me, and gave me hope for a purpose in my life. Because of him, I went to Christian seminary, only to discover that this seminary was a training ground for Spiritual Warriors of the Renegade kind. While the church communities I tried to connect with in order to become an ordained minister were not quite that "advanced" theologically, and not quite as "Renegade" as many of us seminary graduates were, the community of people who I met through my seminary training gave me the basis for the connection I now have with the kinds of Renegade spirituality I am now involved in.

I use the word RENEGADE here partly because Robert Morning Sky uses this term a lot. From my Vision Quest experience I also came to reflect upon my spiritual path, and how I had to leave the church behind, and realized that Chicago Theological Seminary had given me the tools, and the permission, to be able to "let go" of the church itself, in order to actualize Who I Am, and more fully become the Renegade that I am today. It is ironical that the very institution that most people would think was designed to promote the continuation of a dogmatic teaching about Christianity, actually gave me the tools and confidence to break out into new territory, and become the Trail Blazer that I hope I am today. I did not even realize at that time, how ahead of me they were, and how they were teaching me things that I couldn't even realize until now. For, even upon graduation, I was still a bit caught up in a certain mentality about what Christianity meant. I still had some unconscious "Episcopal" theologies rolling around in my head, from studying Morton Kelsey, which I had not yet dealt with. NOW I can see all too clearly that some of my professors were spiritual Renegades, but I could not see it then.

Still, I am glad that I do not need the approval of the Christian communities to do the work I am now doing, and that I do not need to use the Christian language any longer. That language got in the way of the Truth, as it held too many years of mis-understandings within its theology. I am free of that language, and thus much freer to think in a new way.

I also wish to be free of any language or tradition that holds me back from true reality and true freedom. That is probably why I was becoming disillusioned with Lakota-style ceremonies and politics. It was becoming just like the church! The same controlling mechanisms, in new guise. But, I feel that Robert Morning Sky's Spirit teaches me to be free of Tradition, too. Just be free.

From this Vision Quest, untraditional and comic as it sometimes was, I learned a great deal. But more than anything, I learned that each step of the way of one's Path is as important as the other. My past has laid the perfect ground work for my present and my future. For if I had never attended Chicago Theological Seminary, and then later left the church, I would not be able to see through things to their inner core of truth, as I am now seeking to do, and to communicate my insights to the world. I thank God, in what ever language you want to use, for the way my life has unfolded. A true moment of inner peace. (June 25, 1995)

MORE ABOUT "MORNING SKY" AND THE "VISION QUEST" EXPERIENCE:

The day after I wrote the first part of this chapter, my personal life took another weird turn. My boyfriend Don told me that he wanted me to start finding a way that I could move out, and give him his space back! In truth, I had been thinking, myself, that another Winter in his mountain home would be hard for me. I was myself thinking that I needed to develop a new community of friends, with a common purpose, and to not be so isolated socially as I had been during the past Winter, living here, since it's so hard to get out in the Winter. But, I always felt so loyal to Don, and knew that if I moved out again, like I had before, that he would interpret it that our relationship was over, at least as far as any exclusiveness, and start looking for someone else. That somehow didn't feel right, either, as by now I felt a bonding that felt like "family" even though we have not married, and often were going our different ways as far as our creative purposes.

In other words, he would always be a good friend, and an important person and influence on my life, even if I should find another boyfriend to take his place, with whom I had more common interests and purposes. He was not trying to be mean, and he would give me some time to find a way to afford to leave, but it still hurt, and felt like the ending of a significant period of my life. He said, too, "If you find another boyfriend who you like more than me, that's OK, too."

At least I had done my homework in preparation for this shift in our relationship, as my "Vision Quest" in Utah had opened me up to the possibility of finding new friends, and perhaps even a new boyfriend and companion, who would have more similar interests, and a compatible community with mine. My admiration for Robert Morning Sky, and the intense Shamanic energy around him, pulled my heart toward working closely with him, or his group. This was like a new conversion experience.

The "Vision Quest" Robert led us on was in no way "Traditional" as far as the kind of Vision Quests I had previously participated in, and yet I could tell it functioned in a similar way as a catalyst of transformation for my life. As my energy shifted toward a desire for more variety in social contact, and perhaps to even move or travel some to connect with this work, Don must have sensed it was time to give me permission to "let go" of him, and go my own way. I guess I needed this "permission" to leave him again to gain my own space and independence, and perhaps even a new relationship.

I also felt and hoped that Robert would become a friend and teacher. From having had Don in my life, I was healed enough within that I didn't need to be desperate for a companion, either. My Oedipal complex has been resolved, so now I can be friends with men, too, in a new way, if I so desire. I have found my own Power and Authority in certain matters, so that I can hope to become an equal friend to people with whom previously I could only have been dependent and needy.

However, I still need to find a way to survive financially. My next test is to find a way to make money that does not make me a complete servant to the System. To have to go backwards, and have normal routine jobs again, would very possibly kill me. After living in the Paradise of a rent-free mountain home, with a companion who did try to be supportive, most of the time, even though we got on each other's case due to lack of space, I really don't know if I can function in any other setting. So, even as I write this, there is a fist of sadness in the pit of my stomach, and I think Don feels it, too. Robert and his Terra Society will have to be a strong influence on my life in order to replace what I am losing. But, at least Spirit has given me this hope that something new is being born in my life, and I will be taken care of.

Robert has a family, and many burdens in trying to get his work out to the world. He will become hard to reach. So, I cannot depend on only him, either. Yet, I think his Spiritual presence will somehow be there for me when I need it. I have been through much harder isolation and pain than this. After all, I will still be able to see Don, it's just that we now know our relationship had "missing pieces" for both of us. There will be a period of grieving, in any case.

While much of this book has been focused on "information" gathering, this chapter gets me back on track with the personal quest and process of my work. Perhaps that is the basis of it, and the information is only secondary.

I am still processing things that happened, and that we saw, on the Canyonlands "Vision Quest." There are many unanswered questions, because Robert had to leave early. The Rangers at the Canyonlands do not like him being there. I am not clear as to the reason, and suspect that part of it is that he may be revealing information that the Government does not want revealed. Yet, I don't know for sure. Other groups go to the Canyonlands for "Vision Quest" type experiences, and others go to see ancient ruins, so he is being singled out in some way. That is sad, for he does a good service for people in leading his Quests.

On the Canyonlands hike to Druid Arch, we saw some unusual parallel cuts in a dry stream bed, that looked more like it had been a quarry than a stream. The Druid Arch rock itself had a part of it that looked like the Egyptian hieroglyph for "King." At Tower Ruin, we saw a small Anasazi ruin that had a face carved in a near-by cliff wall, which reminded me of the big-eyed Grey aliens. I had a dream in which an Ankh appeared. And there was a very strange storm, with thunder that would not stop. But everyone experienced these things differently. My version and meaning for it would be different than that of others. Out of courtesy to the sacredness of this experience for us all, I won't tell any more here.

As for my process, when the Ankh dream woke me, it started to rain heavily, and I suddenly realized that I needed to tell Robert some of the purposes of my Vision Quest, and the prayers and needs I had to be answered. I sat in my truck and wrote him a note about this. I felt he needed to know more, in order to help the process, and because he had not yet gotten all of the information I had earlier sent to him in the mail, which included copies of my two previous books. In my note to him, I voiced a concern for finding my true "home," as I knew that living with Don had its uncomfortable aspects, and that it still did not feel like a complete "home" for me. There was something missing. I was still searching for my true "home," perhaps from another life time, even.

One evening, when Robert gave us a teaching, he seemed to look particularly at me, and said, "Tomorrow when you see the ruin at Tower Ruin, you will find your 'home.'"

While it wasn't a very big ruin, the fact that I might have seen a portrait of a "Grey alien" carved in the cliff wall near it, causes me to wonder. This was not the same kind of ruin found at Mesa Verde, because I did not see any "faces" there. The energy was different, too.

Then I remembered a dream that I had written about in my **Medicine Story** book, that I had on December 27, 1991. This is the dream: "I had a long dream in which I saw a larger house that had many odd dimensions to it. It had areas where one could climb out on the roof, and climb down via a ladder." (Note: while this dream foreshadowed what Don's house was like, it also seems to be describing an Anasazi type of dwelling, as well -- something like those at Mesa Verde, because larger than the one at Tower Ruin.)

The dream also showed an ancient ruin that included a circle on the ground where an ancient Aztec civilization had held ceremonies. ("Aztec" according to the dream, and reminding me of the large Kiva in Aztec, New Mexico.)

At this circle ruin, I met a woman in the dream. I actually met a similar woman the next day, and we became friends.

And this is the most interesting part of the dream: "The dream ended by showing me a map of the United States, with the path of my journey to Colorado. It gave me perspective on how far I had come from Illinois, and that these "ruins" were very close to me in Colorado, as opposed to Illinois, but I was **not quite at the ruins yet.** The ruins that my spirit seemed to be seeking were farther South West from Denver, and I suspected were closer to the Four Corners area. The dream implied that I was **not quite there yet**--not quite where I was eventually supposed to be. I felt that the dream was telling me that these ancient ruins were "calling" me, like some kind of "homing" signal. They had influenced my move West, but I still didn't feel quite "at home" yet in Denver" (or later, in Idaho Springs.)

Right now, as I am emotionally losing a "home" that has

been a significant part of my journey, I can only hope that I am still being led to find my true "home," and that perhaps this home is with the community of Robert's Terra Society. But, even if it isn't, I pray that I am getting closer all the time.

A dream voice that I mentioned in my first book had said, "You have relatives in Arizona." I suspected it meant Native American clans who I had lived with in previous lives. As most of the members of the Terra Society are from Arizona, and so is Robert, there does seem to have been some premonition to that dream.

I feel relieved that Robert is Hopi and Apache, and not Lakota, and that he is creative enough, and Renegade enough, to be different than the Lakotas. I was getting tired of being told how everything I did was wrong, and of being in the middle of so much spiritual and political turmoil as the Lakota communities seemed to be in constantly. In that way, my Spirit has come home. I can be free to follow my Vision, and be who I am. Ho!

MORE THOUGHTS AND INFORMATION:

Oddly enough, a few days after Don had pronounced that I should consider finding a way to move out, he gradually relented. I told him that if he just let things happen, that I was probably going to move for the Winter, anyway, as I was really getting tired of being cold and isolated all Winter, and driving in dangerous mountain snow and ice. The other part of it was that he had been facing constant set-backs with his building ideas and inventions, combined with my sometimes not being supportive enough. However, after he had a couple fits of anger, and broke some household items to vent it, he gradually calmed. Then Don got several new ideas to try out on his concrete building panels, which boosted his confidence. (After I had prayed for him with my Pipe.) And he soon began some consulting work with a concrete company, starting to make money, after having been living on savings and credit cards for a year!

The other stressor we had been facing was the weird, colder and wetter than usual weather we had been having for months! Today is July 3rd, and it was 40 degrees, and rainy! It's been like this for weeks, and we haven't really had any Summer.

Even on the mountain here it is usually hot and sunny in June and July, especially July. I wore a Winter jacket to town to do laundry today! This "weather warfare," or what ever it is, is contributing to our irritability.

Perhaps I also over-react to the thought of going back out into the world and trying to support myself. But, especially, in the Summer, as I may wish to take a few trips, and jobs get in the way of the necessary time off. And, especially now, as I may wish to be more free to connect with some of the activities of the Terra Society.

So, there seems to be a complete turn-around in our relationship, and even though I still don't know what I will do come Winter, I am not forced to make a move before the time is right. Perhaps some negative force was trying to disrupt my work, as I needed this time for writing, to help get this book finished. It seems my prayers for help were answered.

Then, as if to impress me because I had been so impressed by Indians like Morning Sky, Don reminded me that he is part Cherokee. His grandfather was half Cherokee, so Don is one eighth Cherokee! And, his grandfather was a brilliant inventor, who made a lot of money from his inventions. Don, too, is an inventor.

More bits of information: Recently, my favorite TV show, **The X-Files**, had Part One of the final episode for this season. I kept hoping they would show Part Two, as it ended with Fox Mulder in a flaming box car buried in the desert! This box car was full of alien bodies, with strange vaccination marks on their dried flesh. He had been located by the "Cancer Man," who worked for the Shadow Government, and they wanted him dead, now. They had murdered his father, who also knew some of the secrets of early UFO events, and was involved in some way. His partner, Dana Scully's name had been on the information that was coded in Navaho, so she may have been experimented on during her missing time period.

It drove me crazy not seeing the conclusion to this show. As I don't have TV on the mountain, I need to have someone video tape the show for me, and we have no idea when the conclusion will be shown. It now appears that it won't be until the new Fall '95 TV season. This book will probably go to press before Part Two of that final episode will be shown.

The reason that I am telling about this **X-files** episode is that I found out from a Native American man at the Canyonlands trip that the Government does indeed have a secret program which uses train box cars buried underground, to store information they wish to hide! It is called the Brown Bag Battalion program. These buried box cars have some kind of electronic net over them so they cannot be found easily. If someone is "eliminated," or made to look like they never existed, all their records may go into one of these box cars. Perhaps even bodies go there, as the alien bodies in the box car in the **X-Files** episode.

So, I kept thinking, maybe that last **X-Files** episode got too close to the truth, and was exposing some really nasty information, so somehow they have not been able to show Part Two, due to some kind of Governmental pressure not to. You would think that they would have shown Part Two the following week.

I also really liked the Native American connections in that final episode. "Nothing disappears without a trace," the Native American Elder, who had been a Navaho Code Speaker in WW II, had said. Was he talking about the Anasazi, the bodies of aliens, UFO information? Maybe all three, and more.

The last time I was writing this chapter, the section about feeling badly that Don had asked me to move out, and my emotional processing of that, something unique also happened. As I was writing, I was in a very relaxed stated of mind, letting the thoughts and feelings be expressed. At one moment, I suddenly sensed an Intelligence in the room, or in some kind of mental contact with me, which was very **deep**. I suddenly **knew** that it was the mind of the Star Elder, Bek'Ti! (Or at least, it was a Being just like him.) It was not some big revelation, but a subtle **knowing**. Also as I sensed that it was him, I also sensed that he was **familiar** to me! That sense of familiarity really made me feel even more so that I was not imagining this experience. However, it passed quickly, and now I could not describe any more than that it felt true at the time.

This experience puzzled me because I have been thinking of Bek'Ti as a physical being, who had crashed from a space vehicle years ago. Robert had said that he was still alive, however. And yet, during his teaching Saturday night at the "Vision Quest," Robert had said something that I jotted down in my notes by candlelight. I happened to find this comment in my notes after the

Bek'Ti experience the other night. He had asked in a mysterious way, "Does Bek'Ti communicate with any of us?"

Apparently, Bek'Ti did use telepathy to communicate with Robert's grandfather. Many of the people who have alien encounters say that they communicate with telepathy. So, he may not have actually been in the room, but may have sent his consciousness to be with me, to communicate his presence. And why? He must have wanted me to know that he is real, and that there is a personal connection between us. It's not just a story from years ago.

This experience could have been some sort of Initiation, or even a Test, to see if I would pick up this information intuitively. For me, it took this Star Elder story into another dimension. It meant that the Star Elder was also a Spiritual Person, for he communicated in ways not just physical! And his mind was very deep. By our human standards, this is a kind of a spiritual communication or experience. And that means that, my involvement with Robert and the Star Elder is a spiritual or "theological" experience.

This takes everything one step farther, into the realm where my Vision told me to, "Write theology for Indians." Because now we are talking about ultimate realities, other dimensions, mental and "spiritual" issues. So, in a sense, the Star Elder story is both an actual physical event, and a spiritual, mental event. It reminds me of the concept of Sacrament in the Church. There is a physical and a spiritual component and reality that are intersecting in a Sacrament. Thus, what Robert teaches is a Sacred Story, but more. There is a theology to what he is doing; it is not just an expose' of the Government's cover-up of the alien presence.

Another thing I wanted to share about the "Vision Quest" experience: As it was quite hot the first day, and I was getting sun burned, I was worried about the second day, when we would have to walk at least ten miles to see the Druid Arch. I knew it would defeat our group if we had to walk in the blazing sun, and get heat stroke. So, I prayed the night before for cloud cover the next day, because we were doing an important and sacred thing, but we did not need to suffer terribly for it, as we just needed the information, and to see certain things along the way.

Well, I was delighted when I woke up and it was cloudy! An unusual thick layer of clouds covered the sky most of the way, and even though there were a couple sprinkles of rain, it never amounted to anything. I told Robert I had prayed for the clouds, and he said, "Thank-you!" But, who knows, maybe he also had prayed for clouds, or our Guides were protecting us all. In any case, it was a neat experience for me. I keep hoping I can connect with the Thunder Beings when I most need to for help like this.

Recently, I had been thinking I needed a new Medicine name. I kept getting something with Medicine Wolf, then Star Medicine Wolf. One night, soon after I returned from the Quest, I had a dream in which I saw this name in neon lights, as if like a sign on an old hotel, but added to it was the word, "Red." The "Red" part was perhaps jutting out at a different angle on the sign. Normally, I would have consciously chosen a color like Blue or Silver to be part of this name, like Blue Star Medicine Wolf, or such. But, my unconscious mind wanted me to be "Red Star Medicine Wolf." Afterward, I came to realize that the word RED probably was telling me that I was once again becoming influenced by the Native American part of myself, that I had come close to ignoring. And, again, perhaps I needed the Red or masculine energy now, to balance out my Blue-ness and the passivity that my life had fallen into. But, this is just a symbolic name, to cause me to think consciously of new parts of myself that are emerging. My books will still bear the Medicine Name of Thunder Stone Dream Woman.

In a way, I am glad that I never was given a "real" Indian name. Because now I realize that I would prefer have been honored to have a Name from Robert Morning Sky, and to become part of his Terra Society clan. The personal connection would have been lacking, otherwise. But, only he will know if this is the right thing to do, someday. (July 3, 1995)

(Later, in 2001, I get the name "Morning Star" from another shamanic teacher, instead.)

Workshop with Elisabeth Kubler-Ross,MD.
Photo © by Karen Degenhart.

CHAPTER TEN:

PYRAMIDS ALL OVER THE PLACE

My connection with Egyptian images and symbols goes back to high school, when I was studying occultism, and the Tarot, and often wore a large Ankh around my neck. I did paintings of pyramids frequently.

In July 1977, I met Elisabeth Kubler-Ross for the second time, and was able to see the King Tut Exhibit on July 30th. The day before I met Elisabeth, I had a dream, which went: "Someone has a vision of building pyramids at certain energy centers all over the country, and perhaps even the world, to prevent earthquakes. The whole country is covered with buildings and pyramids, at these special places." The very next day, July 27th, I met with Elisabeth and she reported that she had just put down some money for the land where her first retreat center, Shanti Nilaya, would be, and that she hoped to build a pyramid there, for healing purposes. (I may repeat this later in more detail in another chapter.) The pyramid idea was just getting popular then.

I have heard David Hatcher Childress say, in a taped radio interview, that the word **pyramid** is Greek for "Fire in the Center." Coincidently, I had done a painting in high school in which a pyramid had a fire in the center of it. If I can find this painting, I may use a black an white version of it for the book cover, or as an illustration for this chapter.

Years later, when I was visiting a lady in Aztec, New Mexico, she talked about searching for a place on an Indian reservation in which she had heard there was a flying saucer buried under a pyramid.

I remember David Hatcher Childress saying that one of the largest pyramids in the world is in China, but they don't like to call it that, because it makes people think of the Egyptian pyramids, and to make a possible connection between them. Of course, Mexico has many pyramid ruins, perhaps a bit different from those in Egypt, but pyramids all the same. Many of the mounds in North America are also pyramid-like, but made of dirt, and often with a flat top.

Recently I read a book called, *Atlantis in Wisconsin*, by Frank Joseph, which talks about ancient civilizations in the Illinois and Wisconsin area, and this place called Rock Lake, Wisconsin, which has a huge mound and smaller, pyramid-shaped rock piles, all under water there. He tries to make a connection between both Mexico and Atlantis, saying that the Aztalan ruin area was an out post for these civilizations. There are huge, ancient, remains of copper mines in northern Michigan, and Joseph believes that the Altanteans traveled to that area to mine copper, and then sell it to Egypt and Europe, thus aiding the Bronze Age.

Another connection with pyramid shaped ruins in North America is revealed in the book, *Pyramids of Montauk,* by Preston B. Nichols and Peter Moon. This book is the third in a series that talks about secret experiments in time and consciousness that the authors feel were conducted at the old Fort Hero Base at Montauk, New York.

On page 47 of Pyramids of Montauk, is a copy of a photo that the authors just happened to discover in their researches of the history of Montauk. This photo, taken before 1911, was discovered in a book, Historic Long Island, published in 1911. The book did not explain the photograph, though the photo was captioned, "The Pyramids--Montauk." This is a photo of a pyramid shaped mound about twenty feet tall, which was located on the beach area. It does not look like a natural sand dune, as it is too precise in shape.

What I found surprising is that this mound is very similar to those found at Rock Lake, Wisconsin. There is also an "Atlantis" connection with the Montauk mounds. One of the authors has a friend who lived in the Montauk area a long time, who said that he once took a school field trip out to a field area where his teacher said Indian mounds had once stood. That was in 1943, so it is clear that someone had deliberately razed them to oblivion since then. The authors suspect the military, or some Government cover-up action.

If David Hatcher Childress' story about Egyptian relics in a cave in the Grand Canyon is also true, the reader may be getting the idea that maybe the Government does not want us to know the truth about ancient history? But, why? Is this some kind of "Classified" information? And which government is doing the covering up? The Government we think we have elected, or the

Secret Government? Who is really in charge?

Now, there are even more strange coincidences. Many of the Montauk Indians have the last name of "Pharaoh," the authors discovered. That in itself would make one think of Egypt. The authors add: "Additionally, the geology of Montauk is different from that of Long Island proper. It is very much like an extended mountain that reaches above the sea. Some say it is a remnant of ancient Atlantis that didn't sink and that the Montauk Indians derived from the ancient Atlanteans. The royal family name of Pharaoh is very suggestive of this particular theory." (*Pyramids of Montauk*, Nichols and Moon, Sky Books, New York, pg. 54.)

During their research for this third "Montauk" book, the authors discovered another interesting fact. Our President Thomas Jefferson "was at one time very interested in the native tongue of the Montauk Indians and wanted to preserve the dying language. In fact, the matter was so important to him that he traveled all the way to Long Island in hopes of retrieving it. Jefferson reported that there were only three old Indian women who could speak the language. He dutifully recorded the vocabulary, alphabet, and even composed some sort of dictionary." (Pg. 48.)

The "Montauk" authors wondered why Jefferson would be so intrigued with an almost extinct Indian language. Then it dawned on them: Jefferson was a Freemason. Perhaps he was searching for some kind of ancient cultural information that the mainstream consciousness of his day would not be interested in, that Freemasons "knew" about. Unfortunately, the story continued that he had then lost most of his research, somehow, in the Potomac!

The authors have a friend named Kenn Arthur, who seems to know a lot about Montauk through some secret connections or esoteric means. He told them that "the real story about Montauk is far more bizarre than Preston could ever imagine." (Pg. 48.) They told him about the "Pyramids of Montauk," and Thomas Jefferson's language research. Kenn responded that "the tongue the Montauks spoke was known as Vril, an ancient Atlantean tongue. Ken said that it was a later version of an even more ancient tongue called Enochian, the language of the angels." (Pg. 49.) He went on to add that the name "Pharaoh" was not only used in Egypt, but also in ATLANTIS. Ken also mentioned knowing about a "series of small pyramids" that had once been at Montauk. "They were supposedly white and appeared to be made of brick like objects." They were somehow covered up or taken underground, he said, in other words, hidden.

Another friend of the authors (or of one of the authors, but I couldn't tell which one, from the way it was written) got them into thinking about the locations of pyramids. This friend felt that the Atlanteans knew about a planetary grid system which somehow energetically supports the physical reality of our planet. The energies of such a grid could be tapped into for various purposes. There seemed to be a suggestion that all pyramids on this planet are built in special locations on this "grid" for purposeful reasons, possibly related to some lost technology that used this "energy." The Montauk area is considered to be one of these special areas. Cairo, Egypt is another. Even the pyramids on Mars are related in some way. (Pgs. 52-55.)

I found this information about the pyramids built on special grid lines or sacred spots to be significant, because of my dream years ago that I quoted at the beginning of this chapter. In fact, I did not consciously realize that my dream would connect so closely with what I was about to write. In those days, I did not consciously know about this grid system theory, either.

The authors of *Pyramids of Montauk* again heard from this "Madame X," as they called her, and she told them that she had read a book entitled, Cairo: Biography of a City, by James Aldridge. "This book explained (amongst other things) that Cairo, whose real name is Al-Kahira, means Mars." There is also a legend that the planet Mars ascended on the horizon as the building of the city of Cairo began. The workers there "had a great fear of Mars because according to legend, that was the planet from which people came down and conquered their land, setting up pyramids and the sphinx." (Pg. 55.)

There was also another coincidence. Mars is the Roman God of War, and has always been connected with Martian, war-like energy. The Egyptian God of War was called Monthu-Re or Monthu, whose temple was at Karnak.. Montu is a linguistic phoneme of Montauk.

Nichols and Moon also mentioned a lot about how the Government managed to swindle their sacred land from the Montauk Tribe. Since many of them interbred with Blacks, there came a point when their tribe was declared extinct! This even though a few remaining blood Montauks **know** that they are directly related to the ancient tribe.

The idea that many of the Montauk Indians had interbred with blacks caused me to think back to a dream I had had, which is recorded in my second book. I had a dream that Sun Bear had temporarily become a "Walk-In," as an old black man. A few days after this dream, I received an issue of the magazine he had started when he was alive, "**Wildfire**," in which there was an article about "Black Indians." (I later had a dream that he had moved on to become part of the "Heavenly Host.")

If the Montauk land was on some kind of energy grid, that could explain the Government's desperate need for it, as well as the Montauk Indians' belief in it as Sacred Ground. If this land was actually a remnant of Atlantis, it could also hold more mysteries than we, or the Montauk book writers, yet understand or know about.

I have a hunch that there will be more connections with Mars revealed in the near future. Even as I write this, I remember some information I wanted to include in my book related to the "Pyramids of Mars."

My companion Don has a book from a 1977 printing, which was first published in 1975: *The Guide to Pyramid Energy*, by Bill Kerrell and Kathy Goggin, put out by *Forces*, at 3015 Nebraska Avenue, Santa Monica, CA. 90404 (there at least in 1977.) In this book, one of the early photos of the pyramids on Mars is shown, plus a comment that now seems incongruous, about Carl Sagan. "At Cornell University, Professor Carl Sagan presented a paper to the ROYAL Society of London, which in turn was printed in their "Proceedings," Volume 189. Professor Sagan gave excellent reasoning in his paper, "The Recognition of Extraterrestrial Intelligence," backed in part by the photo presented here." (From pg. 115, Kerrell, and Goggin.) I found this strange to read, as today don't we hear more about how skeptical and conservative Carl Sagan is regarding any "proof" of extraterrestrial intelligence? (By 2010, he has died.)

The authors also suggested that perhaps all our questions about these pyramids on Mars, that do not seem to be of natural origin, would be answered during a 1976 satellite landing on Mars. Instead, today in 1995, after the "malfunction" of the probe that was to take photos in September of 1993, we know even less, and the topic of "Pyramids on Mars" seems to be under some kind of cover-up--excepting the work of Richard Hoagland.

So, now we have some kind of pyramidal connection between ancient Atlantis, Mars, Egypt, Rock Lake, Wisconsin, and even Montauk, New York.

There is also mention of the Monuments of Mars in the book, *Nothing in This Book is True, But It's Exactly How Things Are*, by Bob Frissell, (Frog, Ltd., P.O. Box 12327, Berkeley, CA. 94712, 1994.) This author says that the Martians were the ancestors of the present-day Greys, who built the complex at Cydonia on Mars, as a time/space vehicle, so that they could project themselves into the future, because Mars was losing its atmosphere. (Pg. 149.)

Robert Morning Sky says that civilizations on Mars, Venus, and other planets, were destroyed during an inter-galactic war.

The book, *You Are Becoming A Galactic Human*, also has some information in a chapter on "Earth's Forgotten History." This book says that the pyramid complex on Mars was a temple complex, which was destroyed one million years ago by "dinoid" groups of aliens (probably similar to Reptilians, or related to today's Greys), but that it was a "center of worship for humans!" (Pg. 101.) The "dinoids" were attacking the "galactic humans," who had a colony or civilization there. These same "dinoids" ravished Mars, Venus, remnants of Earth civilizations, and destroyed parts of other planets, as well. (Pg. 101, Essene and Nidle, SEE Publishing Company, 1556 Halford Avenue, #288, Santa Clara, CA. 95051, 1994.)

So, there is some parallel information, and some differences. A few mysteries to resolve about Pyramids on Mars. (7/21/95; 8/28/95)

Don and his light weight concrete panel invention

CHAPTER ELEVEN:

THE WAR IN THE HEAVENS

The idea of a "Spiritual War" is not new, and in fact is a common Biblical theme. However, recent coincidences in my research have led me to think again about this concept. In Robert Morning Sky's book *The Terra Papers*, he tells the story of **Bek'Ti**, a "Star Elder" who was rescued from his crashed disk-shaped craft by six young Native American men, and nursed back to life. The part of this story that I haven't yet mentioned is that the "Star Elder" told the six that "there was a war in the skies above, his ship had been downed by enemy forces." (From *The Terra Papers*, Introduction.)

This is not the only place where I have heard this idea, recently. In a Secret Information Network video called **An Hour with Stan Deyo**, Stan Deyo mentions that he feels there is a kind of spiritual or inter-dimensional war going on, and that in fact, we have probably already been invaded! The key to this invasion is that no one knows it has happened. Yet we know there are events like the JFK assassination which even movies such as JFK reveal to have been a conspiracy effort, and yet no official source ever shows any interest in really getting to the truth. These are ways in which those who have "taken over" the power and financial bases of our planet protect themselves from being discovered. If we can be convinced that there is no war over the control of our planet, then we will not do anything to win it. The biggest tool the Enemy have over us is to control our minds, and the information we receive. Thus, most of the war can be won through simple manipulation of the media.

I think back, too, to my second book in which the "mental patient," who said he was a hit man for the Government, told me "I've seen aliens of all kinds, crustacean-base, reptilian-base, all kinds. The President knows about them. We've got our own ships. (...) There are wars in space. (...) And, the Bible is taken more seriously in **space** than it is here. It talks about wars in the heavens." (From page 84, **Beautiful Simplicity.**) I left out some parts of this quote that I am not **sure** of, or that would confuse my current point.

Another source of this "spiritual war" idea is Dr. Frank Stranges, who says that he's been in contact with a man who is from the inside of Venus, who is very spiritual, and can read his mind, and appear and disappear at will, almost like an angel. This person is called Val Thor, who comes here in a space ship, which Dr. Stranges says he has been on numerous times over the years. Val Thor also is helping to reveal information to Dr. Stranges, who is also a minister, that could help humanity to become free of this mental and spiritual bondage we are under. The details of this story are in the book, *Stranger at the Pentagon*, by D. Frank Stranges, (and currently published by Inner Light Publications, P.O. Box 753, New Brunswick, N.J. 08903, 1991 printing.)

While this story seemed to me at first to be a bit weird, I have now heard Dr. Stranges speak twice, once at the Colorado Springs UFO Expo put on by Timothy Beckley, and only last night (July 22, 1995) at the Denver UFO Society, and feel that he is sincere, and not a "flake." He is also a man of courage, as he stands up to the Government, and demands to know the truth. He also reveals publicly much secret information about what is going on in our galaxy. Though a minister, he's revealing information about the Bible and the Dead Sea Scrolls which truly suggests that alien technology and intervention had a lot to do with Biblical miracles, and the creation of our religious ideas. Yet, this does not ruin his faith; instead his experiences with Val Thor and other spiritually advanced aliens only strengthens his faith in God, and the need to take sides in this spiritual battle we are involved in.

The only thing right now that "bothers me about Dr. Stranges is some of the theological language he uses, and his focus on Christ. That is my personal issue, as I received a Masters of Divinity, and later left the church for Native American spirituality, so I am still sensitive to the potential mis-use of "Christian" language for mind control purposes. I can still appreciate the idea of a spiritual battle, or a battle of consciousness.

Thus, just as in **STAR TREK**, there are good guys and bad guys in space. And, in fact, the information I am getting lately is that there are good and bad guys who can operate inter-dimensionally, too.

From several sources, one being Robert Morning Sky's workshop and **Terra Papers**, and from information shared at UFO meetings, I have come to hear about an ancient Reptilian race which has been in control of our galaxy for a long time. They are very advanced technologically, and in fact they may be responsible for genetically creating humans to use as laborers in their mining operations.

The "Greys" who we hear about are the little aliens with big eyes who are known for abducting humans for genetic material. One article that I picked up at the UFO Society says that the Greys are servants of the Reptilian Masters, and they were also created by the Reptilians. Many of the Greys are coming from Zeta Reticulan, a neighboring star system to Orion. The Greys are usually under four and a half feet tall, greyish silver in color, and have no sexual or digestive parts, as they are created through a cloning and genetic engineering process. They sometimes also have an "insectodial" appearance. The same source says that there are also seven to eight feet tall "Greys" who are from a star system in the Orion constellation called Belletrax. They also look insectoidal. They usually have larger noses. They prefer to influence humans through political control.

Since these cloned Grey aliens have no digestive system, they must take in nutrition through their skin. The easiest way for them to do this is to take a bath or swim in a large vat filled with pureed animal (and sometimes human) parts and glands. This is the reason for the cattle mutilations, as cattle and humans have some genetic and blood similarities. Thus, the Earth is seen by some aliens as a big farm for creating genetic material, and raising humans to "eat." They have been abducting humans for centuries. Naturally, if this is true, it explains why the Secret Government would attempt to make some kind of deal with them not to use humans, except for genetic reproductive material, and instead allow them to take cattle, and allow a "cover-up"regarding the cattle mutilation mystery.

There is also a connection with early religions that involved human sacrifice. The aliens who needed humans to live encouraged these religions. But, it is even more sinister than aliens needing human organs for nourishment.

Much of the above information about Greys and alien nutrition was taken from an issue of "Revelations of Awareness: The Cosmic Newsletter," (P.O. Box 115, Olympia, Washington, 98507. I am not sure which issue, as only parts of it were given out at the UFO meeting.)

A book by George C. Andrews called, *Extra-Terrestrial Friends And Foes*, (Illuminet Press, 1993, P.O. Box 2808, Lilburn, GA. 30226) gives further information about these Grey aliens. On page 199, he gives some information from the Dulce papers, referring to the Dulce Base previously mentioned in this book. "These papers tell what the aliens are after and how the blood (taken from the cows) is used. Aliens seem to absorb atoms to "eat." Aliens put hands in blood, sort of like a sponge, for nourishment. It's not just food they want, the DNA in cattle and humans is being altered." (Pg. 199.)

This same book tells even more horrible information: "One of the allegations concerning the short Greys is that they have a way of deriving nourishment from what leaves our bodies at the moment of death, that they are eaters of souls, who extract from our spirits a certain principle, through a process comparable to the extraction of hemoglobin from blood, which they use as food,...."

"The etheric body is charged with energy that is neither as subtle as that of the soul, nor as coarse as that of the physical body, but is intermediate between them. This is a substance highly valued as nourishment by certain other life-forms, which include the Greys. This is what the Greys are after, which they are here to trap as we trap fur-bearing animals. We are now and have for millennia been used by the Greys as a food source, with almost zero awareness of the actual situation. However, the continued survival of the Greys on Earth depends on our continue ignorance of this symbiotic relationship, in which the benefits are all strictly one-way."

"It is worth repeating that they are using us to keep themselves alive. Fear is the emotion the Greys provoke and feed upon. Once you're afraid, they've got you. Fear puts your heart and mind out of synch, scrambles your energy patterns, and throws your ability to think clearly out of focus, enabling them to take control of you. But you can break this spell by calling out the name of someone you really love, or sincerely respect, or genuinely admire." (Pg. 232, Andrews.)

George C. Andrews continues this information. He says that once a religious system gets established, it loses the truth of the original founder, and becomes a tool of control. The symbols used in this system become tools for a "soul-trap." When the soul leaves the body, it goes toward these familiar religious symbols, where the freshly freed **etheric body** is captured, to provide nourishment for the Greys.

As George C. Andrews writes: "Because of religious conditioning before death, the spirit of the deceased voluntarily approaches a symbol in its path, which is actually a camouflaged instrument that magnetically captures the etheric body, laden with the individual's life-experience. After having been separated from its etheric body, the soul no longer has any memory of its previous lifetime's experience, and thus reincarnates again remembering nothing, as it begins to accumulate yet another body of vital experience. The process goes on and on indefinitely, providing nourishment for the Greys, but paralyzing our evolutionary development by denying us access to the hard-earned wisdom and knowledge gained by experience during our past lives."

"The Greys are other-dimensional predatory parasites, who have been manipulating us, as we manipulate cattle, since time immemorial. Their hypnotic control over us is so complete that hardly anyone even begins to realize that they actually exist. But just as one can take effective means to rid oneself of a tapeworm, when one realizes what the problem is, so humanity now has the opportunity to rid itself of these other-dimensional extra-terrestrial parasites. An ever increasingly widespread realization of the true nature of this predatory game is what will bring about our liberation. The moment each individual realizes that he or she has a direct personal private line to the Source of All, organized religion becomes obsolete and the etheric body eludes capture, because it no longer seeks a symbol or a middle-man, but accompanies the soul straight back to where it came from: God, the Source of All." (Andrews, Pg. 232, 233.)

And more: "By eliminating all references to reincarnation or extra-terrestrials from the *Bible* at the Council of Nicea in 325 A.D., humanity's path to liberation was transformed into a soul-trap almost impossible to escape." (Pg. 234, Andrews.)

The "**Revelations of Awareness"** newsletter goes further into the way some aliens feed off of human souls through **human sacrifice.** "This Awareness indicates the Eucharist, for instance, eating the bread to symbolize the body of Christ: What does this mean? This Awareness indicates that this ritual was that which came down from earlier rituals and it is simply a refection of the aliens eating the bodies of innocents, feeding off of the energies of innocents. This Awareness indicates the phrase 'Food for the Gods,' or 'Food of the Gods,' or 'Nectar of the Gods,' takes on a new meaning when entities understand these things more clearly. This Awareness indicates that indeed, the true nectar of the gods or that which the aliens most prize, is a substance that is taken from freshly-killed humans, which is caused at the moment of death by the strong adrenaline surge."

"This Awareness indicates that the surge of adrenaline through the body accumulates in the bottom of the brain, the brain stem, and these aliens thrive on this substance as though it were some kind of ultimate drug for their particular species. This Awareness indicates that it is most potent in **children**. This Awareness indicates from this, all of the other aberrations of sacrifice of innocents that have been historically present in human behavior, such as throwing the virgin into the volcano, or sacrificing children for some unholy purpose; all of these things, including the present-day Satanic rituals and activities, comes from this particular aberration of these aliens."

George C. Andrews' book also mentions this idea of the aliens' preference for human sacrifice. He says on page 228 that " ... the Aztec priests sacrificed prisoners, cutting out their hearts in a violent ceremony." Reptoids from nearby caverns were given the bodies of the victims, while the Insectoid Grey types sucked the Etheric body released.

Robert Morning Sky has also mentioned in his talks that we are always told to go toward the light after we die. Yet, he feels that "In the light is the re-programming table that erases memories." If we turn around from the first bright light we see, perhaps we will see another light, or find our own light, and avoid this "re-programming" process. After reading the previously quoted information about the entities which "eat" our etheric bodies, which contain the memories, Robert's statements make more sense than they did when I first heard them.

"This Awareness indicates that these reptilian types that are threatening to come here in the future to your planet; these are entities who in ancient times affected much of your history." They are represented by the serpent in the Garden of Eden, which was a laboratory on Mars where the original homo sapiens were created, this source says. (Pg. 8, of an issue of the newsletter mentioned.)

The various Grey aliens are here, this same newsletter says, " ... to enslave humanity, to have humanity work for not just these entities, for they were sent here to prepare the way for the reptilian race." "his Awareness indicated that there is a likelihood or strong possibility that things may come to pass without the masses even knowing what has occurred. Those of you who are aware, may, with careful observation realize what has transpired. This Awareness indicates that there is not something that any of you individually can do or should attempt to do." Just be aware, and use this knowledge to give oneself an edge, more understanding of what is transpiring.

In perfect synchronicity with this discussion about Reptilian Aliens, a friend who I met on the Utah trip with Robert Morning Sky, mailed me some information which arrived today, July 24, 1995. The information he mailed is quoted and discussed below, for it was part of Chapter One of Michael Harner's *The Way of the Shaman.*

In the book, The Way of the Shaman, by Michael Harner, (A Bantam New Age Book from Harper and Row, New York, N.Y., 1980, etc.) in Chapter One, Michael Harner tells of the drug-induced shamanic experience which had inspired him to become a shaman. He was staying in a Conibo Indian village, where he had the chance to take a sacred drink made from ayahuasca, from the cawa plant. Taking this drug brings ones body very near death, and induces visions that only the dying can have.

He then has a vision of giant reptilian creatures who would let him have this information only because he was dying. To quote the book:

"Then they projected a visual scene in front of me. First they showed me the planet Earth as it was eons ago, before there was any life on it. I saw an ocean, barren land, and a bright blue sky. Then black specks dropped from the sky by the hundreds and landed in front of me on the barren landscape. I could see that the "specks" were actually large, shiny, black creatures with stubby pterodactyl-like wings and huge whale-like bodies. Their heads were not visible to me. They flopped down, utterly exhausted from their trip, resting for eons. They explained to me in a kind of thought language that they were fleeing from something out in space. They had come to the planet Earth to escape their enemy."

"The creatures showed me how they had created life on the

planet in order to hide within the multitudinous forms and thus disguise their presence. Before me, the magnificence of plant and animal creation and specialization -- hundreds of millions of years of activity -- took place on a scale and with a vividness impossible to describe. I learned that the dragon-like creatures were thus inside of all forms of life, including man.* They were the true masters of humanity and the entire planet, they told me. We humans were but the receptacles and servants of these creatures. For this reason they could speak to me from within myself."

"*In retrospect one could say they were almost like DNA, although at that time, 1961, I knew nothing of DNA." (Pg. 5,6, Harner.)

Then, Harner feared he was about to die: "I knew I had only a moment to live. Strangely, I had no fear of the bird-headed people; they were welcome to have my soul if they could keep it. But I was afraid that somehow my soul might not remain on the horizontal plane of the fjord but might, through processes unknown but felt and dreaded, be acquired or re-acquired by the dragon-like denizens of the depths."

"I suddenly felt my distinctive humanness, the contrast between my species and the ancient reptilian ancestors. I began to struggle against returning to the ancient ones, who were beginning to feel increasingly alien and possibly evil. Each heart beat was a major understanding. I turned to human help."

"With an unimaginable last effort, I barely managed to utter one word to the Indians: 'Medicine!'"

After this, they managed to make an antidote and forced it into his mouth. As this happened, a powerful being came to help him in his vision, and the reptilian creatures faded back into the lower depths of his mind.

As Michael Harner tried to recall his shamanic vision for his tape recorder, he realized there was a blank part he could not remember. He writes: "The work of recall went easily except for one portion of the trance that I could not remember. It remained blank, as though a tape had been erased. I struggled for hours to remember what had happened in that part of the experience, and I virtually wrestled it back into my consciousness. The recalcitrant material turned out to be the communication from the dragon-like creatures, including the revelation of their role in the evolution of life on this planet and their innate domination of living matter, including man. I was highly excited at rediscovering this material, and could not help but feel that I was not supposed to be able to bring it back from the nether regions of the mind."

"I even had a peculiar sense of fear for my safety, because I now possessed a secret that the creatures had indicated was only intended for the dying." He then went to share his information with some missionaries, who quickly pointed out some passages in Revelations that matched some of his experiences. (Pg. 7,8)

He then went to the most knowledgeable of the tribal shamans, an old blind man, and told his vision story to him. Michael described the giant bat-like dragons, "who said that they were the true masters of the world."

"He stared up toward me with his sightless eyes, and said with a grin, 'Oh, they are always saying that. But they are only the Masters of Outer Darkness.'"

With that, the old man waved toward the sky. Michael realized that he had not yet told the old man that the creatures had come from outer space in his trance vision. He realized that the old man had seen the same thing he had. This caused Michael to decide to learn everything he could about shamanism.

This story supports the information given by the other sources I have quoted, and information given by Robert Morning Sky about Reptilian aliens who created life on this planet. (See **The Terra Papers**.) As of the writing of this chapter, I have not yet had access to his Terra Papers II, or the complete **Story of Bek'Ti.** So, I am only putting together clues from various sources, and still very much a novice regarding Robert's information, even though I instinctively feel he's got something special to offer.

So, this much of the chapter has revealed some information about the Bad Guys, in our Spiritual War.

But, there are Good Guys, too. The book, *The Secret Places of the Lion*, by George Hunt Williamson, talks about advanced souls who reincarnated again and again in prominent positions to positively influence the direction of human history. This book also has some interesting references to ancient history, which coincidently parallel some of what I have already mentioned in my book. Yet, this "Secret Places" book was written in the 50's, published in 1958. The author talks about connections with Star People which are only coming out NOW, in the 90's! Williamson says that he received his information for this book from a Mystery

School high in the Andes Mountains of Peru, near Lake Titicaca. (Pg. 1,27) I will mention more of this later.

In the passages from Michael Harner's *The Way of The Shaman*, he mentioned that perhaps he could access this Reptilian consciousness and ancient information about how they developed life on this planet, because there was some Reptilian DNA inside him, in the lower more primitive parts of his brain.

In the May 1995 "**Discover** magazine, coincidently, there is an article on page 25 which says that "crocodile and human hemoglobin are surprisingly similar." It also states, "Crocodiles and humans diverged from a common ancestor some 300 million years ago,...!"

Robert Morning Sky talks about how a being called Enki helped to create man as a genetic experiment, and that Enki wanted to give this new life free will. An angelic-type of being of a type called the Aku presented him a physical strand of DNA. This was the strand of DNA that gave humans the ability to evolve their free will and their passion. Humans were genetically created from various kinds of DNA, with Monkey and Wolf being predominant sources, according to Morning Sky. However, this one very subtle, almost non-physical (by our standards) gift of DNA from the Aku, which Enki put into his creation of the human, gave us the chance to develop something like Soul, an independent mind, and free will. Robert calls this strand of DNA, "The Gift of the Feather." This is what makes us more than just a beast, but it is what frustrates us when we cannot live fully our passion. Thus, our Soul comes from this strand of DNA, which was a spiritual gift given to us when we were created.

Unfortunately, the Bad Guy aliens seem to feed off of our soul and our passion, and they want to keep us ignorant and in bondage. Some say it is not only the passion (etheric energy?), but also the actual DNA which we carry, that the Grey aliens are after. In any case, we have something they need, "An energy as precious as gold," to them, as Morning Sky says. And we so easily let our guard down, and let them erase our memories, and re-program us.

Soon after I heard these ideas from Morning Sky, and also happened to re-read parts of Extra-Terrestrial Friends And Foes, I discovered, synchronistically, that Mary Summer Rain, in her book Bittersweet, says something very similar about the gift of DNA. Her source for this information is her "Starman" connection. She writes on page 151 in Bittersweet: "Certain physics books spark my attention, like the time I picked up a couple great ones that included The Holographic Universe. That one made me smile as I read it because it hit so some extremely critical facts that my Starman had revealed to me about reality and how all of life is related through the vibrational frequency of one strand of DNA that is God's own essence. Oh yes, I found a few points in this book that the author didn't get just right, but on the whole, he had the germ of the right idea."

This is too much to be a coincidence, as Robert also got his information from a "Star Elder." From this it seems that some "aliens" must be the Good Guys, as they are trying to teach us the truth, and help us to be spiritually free.

I think that I don't fully understand all the cast of characters in this drama, but hopefully this sketchy introduction will give the reader some sense of what may be going on.

See Karen's article "Robert Morning Sky and the Extraterrestrial Connection" in **FATE** Magazine, September 1996, for more info about Robert Morning Sky.

Robert Morning Sky Canyonlands, Utah, 1995

CHAPTER TWELVE:

MORE ABOUT ANCIENT HISTORY

"Man is a 'star' bound to a body! Man is **space born**, a child of the stars. Man is a flame bound to a mountain!" -- Thoth (Pg. 22, Williamson.)

Robert Morning Sky teaches that the Reptilian race evolved into humanoid form before we did. Again, I find the same idea expressed in the book *Extra-Terrestrial Friends and Foes,* by George C. Andrews. He quotes a researcher by the name of Tal Le Vesque extensively. In fact some of the words I quoted in the last chapter turn out to be from this researcher. In any case, it can all be found in Andrews' book. So, Tal Le Vesque writes: "At this point I will tell you a secret: The Reptoids are fellow members of humanity. We are hu (animal) man, and they are reptilian man. They became self-conscious first, before we appeared on the scene. They taught us. However, we both got manipulated and tricked into becoming enemies at war with each other. The astral Insectoids deceived us both, and we are still in a species war with our fellow reptilian man. The real physical danger comes from the adversarial conflict in our struggle with each other. However, the astral and spiritual danger is from the Insectoid mentality, which is alien to both humans and Reptoids." (Pgs. 228, 229, Andrews)

"The Insectoid power fears the Black races and has kept them subjected and confused for they hold the key to knowledge. Their genetic clan information is needed to unlock advanced secrets from the Sirius data. Segmented data units are also carried by Indian clans, some of which were originally from the fourth planet in the Sirius system." (Pg. 229)

"There are cross-breeds, human-looking in form. The Rh-negatives are our link to the reptilian humanoids. They are mostly found among the Celtic people, the English, the Hebrews, certain Indian tribes, and some groups of Black people in Africa. The Reptoids cross-breed with everybody." (Pg. 229)

Tal Le Vesque continues: "About 12,000 years ago a battle for the possession of the surface of the Earth occurred, accompanied by massive destruction and flooding. At that time most but not all of the Elder Race and the Reptoids left the Earth. Their confrontation was broken off until that future time, which now approaches: The Elder Angels verses the Dragon/Serpent beings." (Pg. 227, Andrews)

"The survivors of the ancient battle secured bases to operate from in the caverns that honeycomb the planet. Certain Amerindian and other groups were temporarily taken into protective caverns, but were later returned to the surface. Some humans survived even though they had remained on the surface during the cataclysm." (Pg. 227)

The same researcher also says that the Greys, most of who are controlled by negative astral thought patterns, live "in certain specific cavern areas, midway between the surface of Earth and its interior lands." (Pg. 229) They are facing extinction because they have used up their DNA through constant cloning. This also explains why they abduct people to take sperm and ova samples.

Now, I want to get back to sharing some information presented in the book, *Secret Places of the Lion*, by George Hunt Williamson, (Destiny Books, Rochester, Vermont, 1958, 1989.) The more I read and re-read this book, the more fascinating it gets.

The first thing that struck my attention was this passage: "The earliest records recorded the arrival of "radiant beings from heaven"; these spiritual creatures were the inhabitants of more majestic and enlightened worlds, but man's ignorance caused him to deify such visitors to the "dark red star" -- planet of "sorrow and cataclysm." (Pg. 7.)

This passage struck me because of the term "red star" for Earth. Back a few chapters, I had received a "Medicine Name" of "Red Star Medicine Wolf" from a dream. Yet, the color red actually did not please me, It left a sad or heavy feeling, instead. Now I know why. The Earth is the Red Star, or planet of sorrow. Thus this name is telling me that part of my life purpose is to be here, on this heavy planet, to bring "Medicine," or correction, healing and balance. The kind of Medicine I will use will be "Wolf" influenced. In the Medicine Cards by Jamie Sams, the Wolf is the Teacher and Trail blazer. Through Teaching, and giving out information, as I am doing in my writing, I will bring Medicine to people.

Also, according to Robert Morning Sky, the Creator Enki used both Ape and Wolf DNA to create man. Therefore, we are also part canine. While that may be hard to believe, a woman on the

Utah trip reported that she had seen a science show on TV recently that said that whales may have descended from wolves! In some way, perhaps all life on this planet is genetically connected. While Wolves are an endangered species, they are our relatives, and this may be why the wolf symbol appears everywhere these days, as a compensation for the wrongs we have done to this animal relative.

I have also recently been using the name of "Star Wolf" on the medicine bags that I make and sell.

The book Secret Places of the Lion talks about "fantastic historical treasures which constitute a great legacy for mankind hidden in secret chambers under some of the wonders of the world!" (Pg. 3.) The **lion** symbol represents divine guardianship, which is why it is often seen around doors and gateways. The book is titled *Secret Places of the Lion* because the ancient records, many of which are soon to be discovered, have been protected by "divine guardianship" from the evil powers in the world, so that the day could come when we are ready to understand the Truth. Unfortunately, the evil powers in the world are still trying to keep this information away from us, but hopefully their time is limited.

Many thousands of years ago, during the time of Lemuria, "the first trade ships arrived from the planet Hesperus (Venus). This land is now in the vicinity of the Grand Canyon of Arizona." (Pg. 7.) This information made me think of the cave in the Grand Canyon which was mentioned earlier in this book. I also happen to think of the coincidence that so many of the people I have met recently live in Arizona: Robert Morning Sky, Vance Davis--and Elisabeth Kubler-Ross recently moved there, because of a fire that wiped out everything she owned back in Virginia, and her son lives there. Also, Sedona, Arizona is such a key location for New Age, UFO and Secret Government related information, which is coming out right now.

There is also supposed to be a buried monument or memorial in Arizona, showing where the first "Light Ships" had landed. This will be revealed in the not-too-distant future. (Pg. 8.) When the forces of Evil became too strong, the Venusians then migrated back to Venus, also from this spot.

While our scientists say no life could live on Venus, we still have the story of Val Thor, who supposedly lives inside Venus. Perhaps if there is anything at all to the Hollow Earth theory, other planets also have an inner world, which is more habitable for life. But, I do not yet know for sure the answer to this; I just wanted to point out the coincidences of information, in case it would give us a clue.

The importance of the Great Pyramid and the Sphinx is discussed extensively in this book. "The Pyramid -- the "altar" in the midst of Egypt--was reared by great masters from Atlantis and its construction was supervised by men from other worlds." Williamson writes, "...there are many secret chambers within the Great Pyramid and that its true entrance lies under the silent object that is like a lion and yet like a man -- the Sphinx!" (Pg. 17.) It is said, too, that the "Pyramid was constructed over a natural magnetic anomaly," so that with its cone-like shape, the uncovered sarcophagus in the King's Chamber became a healing chamber. It was never meant to be a tomb, but was always used for purposes of healing or Initiation. These days, the magnetic force is not as strong as it use to be.

Williamson writes, too, that there are sea-shells at the base of the Pyramid, showing that it was constructed before a great flood. This idea is also substantiated today by more recent studies of the erosion around the base of the Sphinx, which some researchers say are signs of **water erosion!** Thus, the Sphinx, too, is suspected to be much older than previously thought. (Pg. 19.)

This book also tells us that there is a most secret chamber near the top of the Great Pyramid which is supposed to contain written records of all world civilizations. The true entrance to the Pyramid was under the Sphinx. "In these underground temples today exist records on clay tablets, tablets of precious metals, and rolls of papyri." (Pg. 20.)

"The capstone was a miniature pyramid itself, made of pure crystalline **copper!** This is the symbolic metal of Venus, and the symbol of Venus is the Crux Ansata (Ankh)--symbol of immortality. It is known that copper possesses a **dimensional quality** unknown in all other earth metals. It was this strange quality that qualified it for the position as capstone. The ancients said, 'Copper is more **precious** than gold!'" (Pg. 20.)

This importance of copper causes me to think about the ancient huge copper mines in Northern Michigan which were talked about in the book, Atlantis in Wisconsin. In that book, the author suggested that the copper was mined by Atlanteans and taken to

Egypt and Europe. Another interesting coincidence. And, in another place this book mentions how the Egyptian army used copper weapons.

Back to Robert Morning Sky: He teaches in The Terra Papers that the Star Being, Bek'Ti, was able to communicate with the Indian youths through using a small crystal that could create images showing the history of the solar system and humanity. It is an interesting coincidence that in *Secret Places of the Lion*, it tells about some "crystal balls originally brought by visitors from outer space during the building of the Great Pyramid. These objects were not solid crystal, but were made up of **nine sections**, fitted and locked together to form a globe or ball. As light passed through them, it was changed in a most unusual manner and the past and future could be partially ascertained in the ball. However, a certain amount of mental control was needed by the user during the period of concentration using the crystal ball." (Pg. 68.)

The rod of Moses in the Old Testament is said here to be an advanced technology left by the space beings, thus, "The orb and scepter of later rulers had their origin in the crystal ball and vril stick--ancient symbols of power!" (68)

Secret Places of the Lion also says that the Valley of the Tombs of the Kings was chosen for its inaccessibility largely because these tombs were not just for Pharaohs, but also to be the hiding places of ancient records. (Pg. 77) Perhaps this idea has been passed on to today's archeologists, for there is talk that the newly discovered tomb of the sons of Ramses II could "uncover many mysteries of the Old Testament, including the story of Moses and the flight of the Israelites from Egypt." (Front page of "**USA TODAY,**" (July 28-30, 1995, article, "Digging Up the Dead."

Again, the *Secret Places of the Lion* says that Ramses II wasn't the evil tyrant that historians and Biblical scholars think he was. Instead, the evil Pharaoh was really Horemheb, who persecuted both Hebrews and Egyptians. He was in league with the Priesthood of Amun, the "bad guy" religious faction, which promoted control through a cult of many gods. (Pgs. 84, 85.) They were opposed to the religion of the One God, or Aton, which had been initiated by Pharaoh Akhnaton. They also murdered King Tut at a young age, to prevent him from resurrecting the religion of Aton, the One God.

Over the years, there were many groups of Israelites who left

Egypt, but the true meaning of the Exodus was that Moses was charged with finding and rescuing sacred writings and knowledge, and getting them out of corrupt Egypt to Israel, where it was now planned for the Christ to be born. This book has some very interesting re-interpretations of Biblical history.

The miracles of the Burning Bush, Moses' Magic Rod, the parting of the Red Sea, and the 'pillar of cloud by day, and pillar of fire at night' that guided the Israelites -- all these were provided by the technology of the space people and space ships! "A gigantic space ship led the Israelites out of Egypt! The 'pillar of a cloud' was the force field, and at night this magnetic field appeared as a 'pillar of fire.' This is exactly how space ships are observed today. Their force fields cause them to become fiery and spectacularly brilliant at night." (Pg. 99)

I seem to remember Dr. Frank Stranges making a similar statement at his recent Denver UFO Society talk. In fact, he says that one of the reasons we have not been given all of the Dead Sea Scrolls that have been translated is because they mention the space beings and how their technology may have influenced and created Biblical miracles! It would upset the religious status-quo, and blow wide open the UFO cover-up.

The Secret Places of the Lion also states that the builders of the Great Pyramid buried one of their space ships near the structure, under "a marker in the form of a lion, yet like unto man." (Pg. 22) This is obviously the Sphinx.

Again, many secret chambers under the Sphinx and the Great Pyramid are to be discovered which tell how Space People came to Egypt, that early Pharaohs were men from other worlds and planets who incarnated on Earth to bring the Light to humanity. There are supposedly even records from other planets stored in these chambers. (Pg. 67.)

There are supposed to be records of antediluvian times now residing "in hidden caverns near the site of the great Pueblo Indian structures." (Pg. 194.) There are said to be "telonium tablets" in caverns, underground temples, and "in buried space ships themselves!" (Pg. 194, 195.) The South West Indians may know of these secret places. They had many legends of Star Beings visiting them, and some of them believe that the "sky gods" "will return soon and take them away with them!" (Pg. 195.)

Again, we find Arizona as being a key location: "The land in

America is known today by the Indian people as "Land of the Sun People." It is also "Land of the Lion." Arizona itself contains the meaning, for "ari in Hebrew means "adult lion." (Pg. 195.)

It is interesting that the *Secret Places of the Lion* suggests this special meaning of the "ari" from Arizona. But, Robert Morning Sky also has an interpretation for that phrase. He says in his **Terra Papers** that the Reptile people called themselves the "ARI" or "Masters" of "AN" or "Heaven." (See Pg. 4) "In time they would be become known as the "ARI" of "AN" (Heaven), or the "ARI-AN." Today it is known as ORION." (Pg. 4.)

Robert further says that the word "ARI" means in Egyptian, "Master or Keeper." (Pg. 5) Of course, the Lion always signified the Royal leadership or power, so perhaps the meanings are similar, after all.

There is also a woman in Australia, named Marilyn Pye, who has been searching for two buried "golden pyramids," which are supposed to be located about ninety miles north of Cairns. She received this clue from a channeling session in 1982. Instead, she found, "a strange basalt crystal rock," which was unearthed from under 24 feet of soil, in Toowoomba! (This is weird to me, because my brother use to live in Toowoomba.) The rock had a symbol of the Sun with "two serpents carved into one side." (Pg. 46, *The American Indian UFO-Starseed Connection*, Timothy Green Beckley, Editor, Inner Light Publications, 1992.)

Marilyn then heard of Hopi prophecies that a white person would bring the symbol of the Sun to the Hopi. She brought the rock to the Hopi Grandfather David, a 100 year old blind man. "He felt the rock's engraving and became very excited. 'This is what we have been waiting for. There are three others to come,' is all he would say." (Pg. 48.)

Marilyn is seeking funding to dig for the "golden pyramids" which she feels exist in Australia, and originate from an ancient civilization long before the time of the flood. Again, in the Secret Places of the Lion it says that the civilization of Lemuria was submerged about twelve thousand years ago. However, some parts that were left above ground were "part of California, Arizona, and other parts of the American South West; Australia, and Easter Island." (pg. 11.)

Again, there are "Pyramids all over the Place" alluding to the possibility of ancient civilizations. And, apparently, too, there are space ships buried under some of them. Or, so the stories say. (August 1, 1995.)

Karen Degenhart with Robert Morning Sky at UFO Expo, 1995

CHAPTER THIRTEEN:

MORE OF MY PROCESS

During the first six weeks after the Utah trip with Robert Morning Sky, I had several emotionally powerful dreams with him in them.

In one I saw him in a Shamanic or Dreamtime consciousness, having on his face aboriginal white face paint, in a grid-like pattern. He then asked, "Do you want me to work with you?"

This dream felt like a true experience, and very profound. I wrote to him immediately, sharing that he could do what ever it took to get me out of my rut, more healed, and actualized for what I need to complete in this life. Thus, he now had my permission to work with me to help me--and perhaps to Teach me, as well. Since we are Free Will beings, it is very important to give another permission to intervene on ones behalf.

Around a week later, still in July 1995, I had a very emotional dream in which I found Robert crouching in a dim room, (perhaps the way he might sit in a Sweat Lodge?) I came over to him, put my arm around him, and said passionately, "Oh, my friend, I've found you!"

Then he put his hand in the air over a spot just below my rib cage, as if to manipulate the energy and release a pent-up past life trauma. I instantly let out a piercing scream, and knew that I had been very suddenly murdered!

For some reason, what later came to mind was that I may have been a sacrificial victim in one of the ancient Aztec heart-tearing-out ceremonies. (What a gruesome thought.)

After this dream, I also "just happened" to pick up my copy of EXTRA-TERRESTRIAL FRIENDS AND FOES, and began to re-read many of the pages I've already quoted and summarized in my Chapter Eleven, which describes some pretty nasty information about the evil aliens who "eat" our Souls, and how they like human sacrifice. From this dream, too, I got the idea to write about the "Spiritual Battle" we are in.

Coincidently, as I am writing this, I remembered that Robert had said something similar during his teaching on the Utah trip.

He had said that the aliens live by sucking our Passion out of us, and that the human body provides the "nectar of the gods," or the Soma they need. This now sounds like something out of some other material I was quoting. Different sources are giving out the same information.

From this same dream I also had a healing glimpse of how my Spirit had been controlled and suppressed for so long -- many life times -- and I felt a connection with a very deep consciousness, much like what I sensed from my brief telepathic encounter with Bek'Ti, the Star Being. It truly seemed as if I had been waiting all my life to meet Robert, and to learn what he had to teach me. Everything that I have done up until now has been preparation for this new Awareness that I am tapping into now, with Robert as catalyst.

These kinds of dreams are still few and fleeting, so I have to use them to help lure me forward on my Path, by remembering their Reality, even during every-day trials, which do their best to crush these insights.

Those few weeks after the Utah trip left me feeling very vulnerable, as it had opened my consciousness, while at the same time, Don was going through moods that made him suggest that we each needed our own space. This opened me up to past feelings of hopelessly struggling to survive, and of emotional abandonment.

But Don calmed down over the next few weeks, and said that he'd never kick me out, unless I had another place to go, and a way to support myself. Still, I began to realize how hard it was to do my writing at his house. I could not use my computer without risk of it shutting off, when ever he was out in the work room drilling, or playing his loud rock and roll music (which I could not stand.) The solar power we had would fluctuate and make the screen dance around, or even turn it off. So, I could rarely do my life work of writing, as he was almost always home, these days.

(Actually, as I write this, I may get a temporary reprieve. Don plans to be doing some "concrete engineering consulting work" for some time in August. That could give me just enough time to finish this book, before I decide about moving, or any other changes.)

And, we'd also fight over the use of the phone, as I had to stay "on line" for many hours to make even a minimal amount of $20. to $25. a day, with $30. or more being a good day. (But there are periods when I don't make anything at all, too.) As this is contract work, I also have to figure out how to pay my own taxes. I tried to get my own private line, but US West had just started giving out private lines in this mountain area, and still didn't have enough for us to have two here. (Later, they did.) With Don getting a Fax machine soon, this will add to the competition for the use of the phone line. People usually take for granted having constant electricity, and phone lines, not to mention heat when its cold out. I have none of these things here, on the mountain.

During July, I had been waiting eagerly for Robert Morning Sky to come to Denver and do a workshop here, and in Boulder. This became a weird experience as well. First his motor home broke down, so he could not make it on time to Denver, then, when he tried to fly in for a workshop in Boulder the next day, the plane's engine would not start, and he never left the runway. Thus, something was keeping Robert away from Colorado, and also from me. I guess there are things I need to deal with by myself. I only pray that my writing about him here in my continued "Medicine Story" does not cause any misunderstanding between us. Obviously, what I am sharing here is all related very intensely to my own Awakening and process. To leave out the "Morning Sky connection" would be to leave out a most significant part of my life. I could not do that and continue this Work. It is all connected.

On July 26th I had this dream: "I was taken to some kind of **Detention Center**, where I was held as a prisoner, with few rights. I was accused of something that I didn't understand or know about. I knew I was innocent, yet they kept treating me like I was guilty of something. They had all my ID's, check book, money, address book, and even my Sacred Pipe! They were keeping my Sacred Pipe as a kind of security, so that I would not run away, as I could get permission to go out for short walks in the area. I thought of running away, but decided to wait and see what was up, as I had no access to my money, or anything I would need. I later saw another woman returning from her "hearing," with her lady lawyer, and crying, for she, too, was accused, and had no idea how to defend herself. It all seemed to be a set up, with no way out. I had something to read for a while, but soon would be totally wasting my time to be there. I also realized that every day I was held there, I could not work on the psychic line, and would be making no money. I didn't have a phone that I could consistently use, or my Tarot cards. It was turning into a nightmare."

In a way, this reminds me of my current situation. Perhaps time is running out for me to be here on the mountain.

Another very unusual dream: "I am hiking to Druid Arch again at Canyonlands, and it's now cold out, and I'm wearing my Winter coat. I get to a spot within two miles of Druid Arch, and I come to a "French" Airport. (How I knew it was French, was that supposedly some French words were on a building there. But, really it was more just suggesting the idea of "foreign" to me.) I meet a guard at a fenced-in area, and manage to say that I'm English, or English speaking, so that he understands. He sees that I am a weary traveler, and offered to bring me a free meal, but I had to stay outside the gate and fenced-in area. The guard said that it was lucky he understood me correctly, or I would have been in trouble (as a spy?) I had the feeling that the "French" idea meant that this airport was in another dimension or time."

This dream reminded me of all the French spoken by researchers in the movie **CLOSE ENCOUNTERS OF THE THIRD KIND**. That movie suggested there was an international secret organization which was in contact with Extra-terrestrials, and that this secret was kept from the public, even from those gifted individuals who were sensing something going on, or should have a "need to know," as the lady whose son had been abducted.

Then, I realized that I had wanted to make some mention in this book about the scene at the beginning of the "Close Encounters" movie, where the airplane pilots who had vanished in the Bermuda Triangle in 1945, suddenly appeared in the desert, years later! There was also the mast of a ship in that scene, as if other ships that had been lost in the Bermuda Triangle had found their way to this spot, as well. And, that reminded me of the movie, **THE PHILADELPHIA EXPERIMENT**, in which a ship had been transported ahead into time during a Naval experiment in 1943.

The key there is Time. In both *The Bermuda Triangle*, and in *Without a Trace*, both by Charles Berlitz, there is an account of a lady pilot named Carolyn Coscio, who had vanished under very unusual circumstances. I will quote from WITHOUT A TRACE, the condensed version of the story:

"Carolyn Coscio, accompanied by a passenger, Richard Rosen, was, according to her flight plan, due to refuel at Grand Turk

Island on her way from Pompano Beach through the Bahamas to Jamaica. At about the same time she should have reached Turk in her Cessna 172, personnel of the small airport could see a Cessna 172 circle the island, could hear the voice of the pilot talking to her passenger, but could not make radio contact from the tower. The pilot observed that a wrong turn must have been taken, that the island was not Turk because, "there's nothing down there." Then she was heard to express concern about the gas supply and to decide to fly to another island, followed by the curious question: "Is there no way out of this?"

"Nothing was ever found of Carolyn Coscio or the Cessna 172. It is assumed that the plane over Turk Island was hers even though it was not definitely identified. But the question remains: If the observers on Turk Island could see the plane and hear the conversation, why couldn't Carolyn Coscio hear the tower? Why couldn't she see the field and buildings on Turk? Could the explanation perhaps be that, through a singular inversion of time, she was, in effect, looking at Turk Island -- but at a moment in time **before** the airport was built?"

"Perhaps "time faults" (if at all possible in the first place) are more common than it would seem." (Pg. 127, 128, *Without a Trace*, Charles Berlitz, Doubleday and Company, N.Y., 1977.)

These Charles Berlitz books are the ones I traded in Sedona from the Indian man who wanted my books in exchange.

And there is more. Last week, I received a copy of a newspaper article from Moab, from the Utah trip organizers. This article, from **"The Times-Independent"** Thursday June 22, 1995, tells about a pilot who was killed the very night of our "unusual storm." The pilot had apparently missed seeing the Canyonlands Field Airport because the "radio-controlled switch that allows airborne pilots to turn them on was not working." Although in this case, the people on the ground were able to speak briefly with the pilot, and manually switched on the lights. By then, he had flown past the airport, and never touched down on the runway. On Monday the downed plane was found a mile from the runway, and the pilot, forty seven year old Christopher Miller of Colorado Springs, was dead from the crash.

As I had my dream of a "French Airport" at Canyon-lands a few days after receiving this article, I feel that my unconscious mind used this "airport" idea to give me this dream to make a link between that unusual storm, and the kind of time and dimensional portals that are sometimes said to be part of the UFO, ET technology phenomena. I hope you can see the connections I am trying to make.

The more I think about how I was awakened up with a dream about an "Ankh"-- in fact an Ankh with an "eternity symbol for a bow tie or cross piece" -- just at the outbreak of the rain part of that storm, the more I felt that I had been **protected** by that Ankh by some higher consciousness. The Ankh is the symbol of Eternal Life to the Egyptians, and I use to wear one constantly during high school. I was surprised to notice, also, how important the Ankh was throughout the book, *Secret Places of The Lion*, which I have recently extensively quoted.

Because of that dream, I also began to communicate to Robert, mostly through notes and letters, of my insights, needs, and dreams. I began to ask for his spiritual help. I began to connect with him, emotionally. Yet, in truth, most of my process with him has been internal. I have so far received few outer responses, clues, or verifications from him as to whether or not I am on the right track with my insights and research. This appears to be a Path that one must tread alone, to much extent. Yet, his presence in my life has been teaching me much. (August 3, 1995.)

CHAPTER FOURTEEN:

I ALREADY HAD THE "KEY": BRINGERS OF THE DAWN

During the Utah trip, someone had mentioned that Robert liked, or agreed with, Barbara Marciniak's book *Bringers of the Dawn.* I had read this book last year, but obviously there is such a shift of consciousness, and so much information in it, that I had already forgotten much of it. So, I took a second look.

I know that when I first started reading this material, it seemed somewhat strange to my intellectual mind. However, some ideas in it were enough to spark new shifts of awareness for me. Some of these idea seeds were planted in my mind, to blossom further after my meeting with Morning Sky. Right now, I can only highly recommend getting, and studying, this book. I will quote or summarize a few ideas that really stand out to me, but I'd have to quote the whole book to do her ideas justice, (and I don't think that her publisher would like that!)

One key idea that Robert had said was that, "Our Key out of here is our Passion." Those of us who are tired of being used as slaves for the System, of having our past life memories taken away from us, and of reincarnating whether we want to or not, are seeking a Key of freeing ourselves from the rut we are in. Passion is also called "Light and Fire," by Robert. It is this Divine Energy that our one special, more spiritual, DNA strand gives us--and it is this Divine Energy that the Grey aliens feed off of. It is also channeled to them through any powerful emotion. Apparently, our bodies are the "containers" of this emotional energy that they take as nourishment.

My own sense is that adrenalin has something to do with it. This was mentioned earlier in my chapter on the Spiritual War. But, since then, I've had personal experiences of becoming over-wrought over emotional issues, and realized that the adrenalin peak that was occurring whenever I became distraught and upset, was probably just what the Greys were after! I began to feel that there were things in my environment, such as Don's moodiness for example, that could increase my anxiety and create more adrenalin in my body. Thus, perhaps I and he were being manipulated by unseen forces just so that the Greys could have a nice meal, periodically, off of us.

I recently visited my Hopi/Mexican Teacher friend, Mr. B., (also know as "The Corn Farmer,") and he, too, said that the Greys, who operate in another dimension, feed off of strong emotion. They come from the Zeta Reticuli binary star system, as George Andrews reveals on page 189 of his book. The Government even has a secret code name for them, which Mr. B mentioned, but I cannot reveal any more details, as his life would be in danger. He was in the Service, as I mentioned earlier in this book, and he had some kind of security clearance. If he were to go public with information that he knows about UFO related stuff, he says that the Government would kill not only him, but everyone he knows! And that would probably include me, as well.

I also shared with Mr. B some of Morning Sky's information about the Star Being rescued by the six young Indians. Mr. B said that the story was true, as one of his relatives also knew about it. In fact, he said that his "great uncle" had been "one of the six!" That surprised me. Mr. B said that nothing was written down, but that the story of the rescued Alien Being had been handed down to him, especially, because there were many omens when he was born that he would be "different." Mr. B did verify that the Alien Being, who was just an ordinary man of his race, did live quite a while, and taught the people things. The Indians kept it quiet, so the Government wouldn't find out. And, the Being eventually died, he said.

The idea that the Star Elder may have died did not match with my experience of the Star Being telepathically connecting with me, unless he did so from the spirit world. Robert seems to suggest that Bek'Ti still communicates with him. Robert, also, does not believe that an Indian of Mr. B's relative's name was among the six. (I must have written him, asking about it.) However, there are enough similarities and coincidences here that I feel the basic story could be true, even if some of the details are unverified. As there are some Dimensional factors here that I may not yet understand, perhaps any seeming contradictions can be resolved in a way that I could not predict.

What seemed most important here was the confirmation from Mr. B (The Corn Farmer) of a similar story, with what Robert

Morning Sky is talking about. Perhaps after all these years, some of the details have been lost. 1947, when the "Star Elder" was supposed to have crashed, is 48 years ago, in 1995! It is easy for some of the details to get lost, or new ones added, by the story tellers, in that amount of time.

Back to Barbara Marciniak. As Robert says, "Our Key out of here is our Passion," so does Barbara say almost the same thing, "Emotion is the Key to getting off this planet; it is the key to figuring out the multidimensional self, healing it, and becoming one." (Pg. 207, *Bringers of the Dawn,* Barbara Marciniak, Bear and Company, 1992.)

As I am writing this, I flipped to another page in Barbara's book that fits with what I have been talking about regarding passion and emotions: "You have been raped of your life force. If there is anything that you as members of the human species have in common it is that you have been raped for your emotions. Others have played your emotions as though they were instruments, and they have never let you know the power you have with your emotions."

"Always this whole story comes back to emotions. Emotions are like tickets that can get you places and plug you in. You are incredibly rich. If you would only realize how wealthy you are with your emotions. The lower vibratory beings, if we may be so bold as to call them that, exist off emotions in a very small range of frequency -- emotions that are based on fear, chaos, and violence." (Marciniak, Pg. 94, 95.)

"Earth's owners have not wanted humans to understand that their feelings and emotions are like a crop, such as wheat, that can be harvested. If you are in charge of your own harvest, then others cannot take advantage of you and use you unless you decree it. When you operate with a certain frequency and sovereignty, those who wish to control you are not interested in you. They want a fearful, chaotic frequency, which is what nourishes them. Fear and chaos have predominated on this planet because these entities have stirred them up. They have divided and conquered everywhere to create that frequency. When you operate in peace and love and with information, you alter the structure of this place drastically: you bring choice of frequency back to this planet." (Pg. 95, 96.)

Now I must pause to say that these things that I have been

quoting or summarizing in this book are not just strange facts outside of my own experience. They perhaps once were, but now I am being given Teachings and experiences constantly which are examples of their Reality, at least in the arena of Consciousness -- which may turn out to be the only arena that matters.

For example, as I became aware of these entities that feed off of intense, especially the lower vibrational, emotions, I would find myself creating these emotions as a matter of habit, using my companion Don as an easy catalyst.

Other times, it seemed that I would tap into old feelings from past times in my life, and still be carrying them around in my vibrational space. This was evidenced by having an unusual depressed feeling several days in a row, if I woke up between 3:30 a.m. to 4:30 a. m. When I had this feeling, it did not seem appropriate for my current reality. It was kind of a hopeless feeling, similar to what I would get when I had to get up early for a job that I didn't like, and felt trapped in, while just barely surviving financially. I may have been anticipating having this kind of experience, again, in the future. Yet, it still didn't feel appropriate for my usual self right now. I remembered reading somewhere about how the Secret Government was often sending microwaves or ELF waves over the population, especially around 3:30 to 4:30 a.m., when many were in a lighter and more suggestible sleep pattern, as a form of "frequency" or mind control. This sounds completely paranoid, but if you do any serious research on it, you will find some surprising information. Broadcasting certain frequencies can cause people to feel depressed, tired, confused, or even happy. I've heard speakers at the Global Sciences group talk about this. It sounds incredible. But, many scientists know about it. And, who is in charge of something like this? An invisible Enemy, one that may have us under control, without us even suspecting it. They control us through controlling our brain waves, our mental and spiritual frequencies, and especially through the Media. This is the kind of thing Barbara Marciniak is telling us about, too. Yet, she blends the scary stuff in with hopeful ways to fight it. Read her book and you will see. (By 2010, we know that the Government and military have PsyOps that use these mind waves in war, read THE MEN WHO STARE AT GOATS, by Jon Ronson.)

So, I suspected that maybe those few days when I woke up and felt depressed, and also couldn't sleep for a while, that I was the subject of some "depression" microwaves, or ELF waves, or some other kind of mental experiment. I don't know if they were for the general population, or just for me, but it made me think.

The next night, strangely enough, I had a dream sensation that there was a "cotton cloud" between me and any harmful energies. It felt as if prayers and helpful energies were around me, protecting me. This time, the words **"Thought Police"** came to me during my sleep, and I wondered if the "Thought Police" were those entities or technologies that could "control" or effect our thoughts via "frequency control" -- just as Barbara Marciniak talks about.

I then visited Mr. B again, and mentioned my unusual depressed feelings, and the words "Thought Police" in my dream. He said the same thing that I had already heard of about the microwaves (or ELF waves), and mentioned a building in Denver where there was some kind of broadcasting station for these thought control waves. You see, if the Greys feed off of low frequency emotions, then they would want us to stay in those frequencies. They would infiltrate the Secret Government and purposely do what ever they could to create more chaos and tension in society. They would want us to stay stuck in the problems we already have. They would even want to create more stress for us. This book may not be the forum that can "prove" any of this, but it is a place where I can introduce the idea to more people, at least.

Yet, you watch, Mary Summer Rain is going to be coming out with similar information in her next book after *BITTERSWEET*, for her "Starman" contact is revealing this same kind of information to her that Robert Morning Sky, and people like Mr. B already know. This may be why she has to have an armed guard around her cabin 24 hours a day. It's not just because she writes about her love of Nature, and her prayers with her Pipe, you can bet! She has had ET contact, and she's planning on writing more about it. Just what the Secret Government fears.

Now, about frequency control. Keeping ones frequency at a certain "higher" level seems to be a way of having Protection around ones self. And, I had that dream of an Ankh at the Canyonlands, during that weird storm. Barbara Marciniak also talks about the Ankh!:

"In ancient Egypt, the ankh, or symbol that represented life, in actuality was a frequency modulator. It was utilized by individuals who were masters of keeping frequencies and able to do many things with those frequencies. The ankh is similar to a tuning fork and can direct sound. That is how it was used a long time ago." (Pg. 192) Sound, is of course, a frequency, so that makes sense that it could be used as a tool of consciousness, and more.

HISTORY IS THE KEY:

Again, as this book started with the idea of searching for the missing pieces of ancient history, so I am led to what Barbara Marciniak (channeled from "the Pleiadians") says of ancient history.

Creator Gods who were master geneticists had a Master Plan for the Earth, to make it into a kind of Living Library of genetic information. Thus, "Many sentient civilizations gave of their DNA to have representation of their coding upon this planet." (Pg. 14.) These Original Planners and Creator Gods were workers for the Light, which is the vehicle of free information and higher consciousness. The Earth would have been a Library or Information Center for many galaxies. It would have been beautiful and peaceful -- a Paradise.

She says that there were civilizations far older than Atlantis or Lemuria, some of which may be buried under the ice caps of Antarctica.

However, the it seems Universe must contain some "bad guys" who wanted to spoil the plan for Earth to be a Living Library and a Paradise, for there were wars in space over the control of planet Earth. "Skirmishes took place, and Earth became a place of duality. Certain creator gods who had the right to do whatever they wanted --because Earth is a free-will zone -- came in and took over. We call this "raiding" the Earth." (Pg. 14, 15) This occurred around 300,000 years ago -- a time that we are taught was the beginning of human civilization, but in reality, it was only the time when this "take over" generated a new phase of human civilization.

After this war in space, a new group of creator gods took control of Earth. "These new owners did not want the native Earth species -- the humans -- to be informed of what took place.

Uninformed, the species would be easier to control." (Pg. 15.) (Humans had already been genetically created by the first, "good guy" creator gods.)

The new Controllers were also master geneticists, and wanted to create or manipulate the life on Earth to suit their needs. They apparently feed off of the "electromagnetic energies of consciousness" which are given off from all life. They had a "different appetite" than the first creator entities. The Controllers needed electro-magnetic energies, or life energies, that came from chaos and fear. "These things fed them, stimulated them, and kept them in power." (Pg. 16.) They came and created early Earth civilizations, such as in Sumaria and Babylon. But, they also created the "Fall" of humanity. They deliberately genetically re-arranged the DNA of the native human species, so that we would "broadcast within a certain limited frequency band whose frequency could feed them and keep them in power." (Pg. 16.) Thus, although humans at one time had access to "Twelve Strands" of DNA, they now can only use "Two Strands."

By coincidence, Robert Morning Sky taught us that we **already had** twelve strands of DNA, and that **one** of them was this finer, more "spiritual" DNA, that was a gift from the original creator gods--the "good guys." He called this DNA "The Gift of the Feather." The other ten strands of DNA are too fine for our current instruments to see, plus they are still in dis-array.

However, the time is coming now that the "good guys," the Original Planners of Light are going to be making significant moves to recapture their territory. They do this by sending spirits to incarnate and "infiltrate the project," who will be able to hold the higher vibrations of Light. (Pg. 17.) The Original Planners are going to change the frequency of the planet to one of Love, thus the food source of the Controllers or "bad guys" will be eliminated. As this happens, our DNA will again evolve into a Twelve Strand system, as we mutate into the higher beings we were meant to be.

Many of us here now are part of this "Family of Light," in disguise as humans. We are here to be **Renegades** against this current negative system, and to be the "holders" of the new Love frequency being sent to Earth. Apparently, the Love energy or frequency could not hold here without living entities who could channel or hold it in their physical bodies, vibrationally. (Pg. 18.)

In a way, the Christ was the forerunner of this new "invasion" by the "good guys." The Entity who was able to "hold" the first strong Love energy, and Light consciousness was the first well-known Renegade to incarnate, after the take over by the "bad guys." (This is my own insight.)

Thus all the theological mumbo jumbo about the Incarnation of God into Man does have importance, after all. But, it is not in the limited way we have been trained, probably by Controllers, into thinking about it. We did need a strong Entity to come into the Earth and begin to hold the infusion of Love and Light that the Original Creators would beam into the Earth, to try to win it back. Who was this courageous Entity? Perhaps he was one of the Original Creators themselves, a kind of God to our limited consciousness. Perhaps Mary had a "Virgin Birth" because Jesus was a genetic implant, designed to give the incarnating Entity a vehicle that could better hold the higher consciousness he would bring into the world.

My gosh! I might have converted myself back to Christianity! I never know what will happen when I sit down at my computer to write. It is as I learned at Seminary: Christ was a Renegade against the system of his day. It is just in the application of His message that we get trapped again, by the Controllers, by Systems, and Institutions--and by our own self-serving interests.

Just as Barbara Marciniak uses the term "Renegade" to describe the Light Workers who have incarnated to infiltrate and "bust" the current system, so does Robert Morning Sky frequently use the term. I don't know if either of them were influenced by the other, but it is an interesting coincidence. Robert calls his "followers," or those who follow his path, Renegades. In this case, the Renegades are the Systems Busters, and the Good Guys -- the Light Workers. It is important to remember that often the Good Guys may appear like the Bad Guys, and the Bad Guys might appear like the Good Guys -- like an Angel of Light. So, Scripture isn't so far off, after all. It's always a matter of interpretation and application, coupled with discernment, and purity of motive.

Barbara Marciniak says, "You are encoded, and as your memory begins to rise, you will respond to the plan with which you came here to participate to alter the frequencies. You will begin to hold, keep, and maintain a certain frequency and then to **live it.**" Our frequency effects everything around us. Many of us already know their assignment, and there are others "whose memories are just beginning to rise." (Pg. 19.)

As I write this, it blows my mind! Each chapter I write is an Initiation into new insights and levels of consciousness.

There is yet another connection between Robert's and Barbara's teachings. Robert talks about how the Reptilian humanoids evolved first, and have been controlling the galaxy for a long time. Barbara also talks about the reptilian space beings, as a race of creator gods. Among them, there are those who wish to control man, and those who are more benevolent.

In another source, I was reading about how scientists think that birds evolved from reptiles. Barbara Marciniak also says that there are birdlike creator gods, insectlike ones, and so on -- in addition to the reptiles. (Pg. 31.) This caused me to think about Robert's teaching of "The Gift of the Feather." If birds, the bearers of feathers, are ancestors of reptiles, then there is another connection between the benevolent DNA gift he speaks of, and the more ancient ancestors, the reptiles. So, there are "good guys" and "bad guys" everywhere, among all races of humanoids. One of the creator gods that wanted to give humans more "passion" and individuality, gave of his own DNA--which, I am guessing now, may have had this reptilian / birdlike DNA--thus Robert's "Gift of the Feather."

(For Robert's sake, I hope that I am getting this right; it has been difficult getting a full account of Robert's teachings, for various reasons. I keep feeling like there is some part missing when I go to put it all together--which is partly why I need to process and put it together in my writings, for my own understanding.)

But Robert had written to me that his purpose was "to convey a message", not to focus on or glorify **self.** Until I realized what the "message" was, which really hit home as I wrote this chapter, I was tempted to become confused and dismayed when ever some person or situation would try to cast a dim light upon some aspect of Robert's character or teachings. Now, I realize that Robert's just a "normal" human being, like we all are -- even as Angelic Renegades -- and any quirks he may have cannot harm the Truth of the Message. Thus, if at times all the details about the story of Bek'Ti don't fit, because different people have slightly different versions of the story -- it doesn't matter. For the Message is what

I have written about here today: We must be Renegades -- the Light Workers who will bring a higher frequency of Love and Light to the Earth. And we must find ways to not be brainwashed and controlled by the System. This is a subtle business of discernment, but we came because we knew we were up to it. A Ho!

My apologies to Robert: It only seemed fair to use his name and give him credit for the Teachings I am discussing, even though he would rather not focus on himself. In this world, there are vehicles of the Message, and he is one of them. In order for more people to access his Message, they will need to know his name, and an address from which to order materials, or find out how to learn more about his Renegade path: Robert Morning Sky, 4727 E. Bell Road, Suite 45-311, Phoenix, AZ. 85032. (It would be helpful to say that you heard about his work in Karen Degenhart's book. Then he will know how much my writings are helping him.)

(Robert had kind of disappeared while George W. Bush was in office, but I heard rumors he may be resurfacing again. However, I am not sure of the above address now in 2010, or for sure where he is now.)

And, I thank Barbara Marciniak for her channeling work that helped it to "all make sense" for me.
(Aug. 8, 1995)

Rock formations in Canyonlands, Utah

CHAPTER FIFTEEN:

THE GIFT--OF BLACK AND WHITE

As I left on my way to Utah, I found a small black feather, probably crow. It fit perfectly on this rugged looking brown Medicine Bag that I was bringing along.

Now, as I am nearing completion of this book, I have come to a point of no return -- a healing point -- in my process with that journey, and with my connection with Robert. It is **a point of no return** because there is no way that I can ever see things the same again, after having connected with his work. I cannot leave him out of my work, either, even if he's uncomfortable with what I am doing. So, today, I found a white feather, right here in the house. It's probably a feather from one of my craft bags, but I don't know how it fell out, or which bag it came out of, or **why** just today it would manifest itself. I felt that it was an answer -- that there was still peace between us, and more healing to come. Perhaps, as I was doing my writing, it was also a sign that I was following my Passion -- my "Gift of the Feather."

The Black and the White, the opposites: The two feathers mark the beginning and the ending of a phase of my process during and after my Utah "Vision Quest." I am still, weeks later, finding my Vision from that Quest. Writing this book helped me to realize it. The beginning and the end . . . .

The completion of a phase of the journey entails a review of the beginning of that journey, and part of mine began years ago, when I connected with Dr. Elisabeth Kubler-Ross. During this time, in a dream, Spirit promised me a Gift -- a Gift of Black and White. During this time, I touched upon both the Egyptian and Native American connections that were to blossom much later in my life. Let me explain . . . .

In October, 1976, the year I graduated with my BA. in English, I managed to get a ride up from the South Suburbs of Chicago to an Evanston Jung Society meeting with Elisabeth Kubler-Ross, MD. I really didn't know much about her work then, but had found out she lived in near-by Flossmoor, and would be speaking at the Jung Institute. It was a long drive to Evanston from

Hazel Crest, and I wasn't sure that I wanted to risk it in my old Rambler, so I was looking for a ride.

It ended up that a priest who was going to ride with her could not make it, so there was room for me to go along. Elisabeth wasn't driving, either, but an architect and his wife were helping out Elisabeth by driving her to her talk. This was October 3, 1976.

This first exposure to Elisabeth had a powerful impact on me. Her talk spoke of life after death visions, and spirits who could materialize and leave notes in their handwriting to prove it. Also, emotional stories about people who were greeted by already dead relatives at their death bed. And, of course, the Four Stages of Dying. And of angels and spirit guides. All things that had been considered taboo for anyone in the medical establishment to discuss.

It amazed me that an "Establishment doctor" would be open to the kind of spiritualistic thought I had already been involved in since high school. I really admired her for coming out with these things.

There was also something about her energy, or the energy around her. It was deep and powerful. As we drove, I remember her pointing out the beauty of the sunset -- referring also to the beauty of the end of life.

On the way back from her talk, Elisabeth had shared about some "miracles" that had happened to her recently. Some spirit guides had actually materialized and communicated with her!

I then shared with her how my grandfather Degenhart had come to me in a dream, after he had died, and he said, "I love you," and told me that I could have his Masonic ring--if I wanted it. He had been a 33rd Degree Mason, and to me this symbolized that he had been a student of the Mysteries. While my Dad thought the ring would be buried with him, a quick phone call to my Grandmother confirmed that she still had the ring, in a box, somewhere -- just as it was in the dream. (However, I didn't get the ring after all. It was given to my Dad, and he, being Lutheran, had it changed from a Masonic emblem to a red ruby, and then when he died in 1996, the ring was given to his brother in law, re his second wife, with the ruby.)

After I told this story about my Grandfather Degenhart during our ride, the man driving said with surprise that he had known my Grandfather! For my Grandfather was a contractor and cabinet maker -- the owner of Harvey Millwork Company (which is no longer in business, now.) The architect hadn't known that my Grandfather belonged to the Masonic Lodge.

I knew from this dream that I had a real message from Grandfather, for the "I love you," message seemed appropriate, and the idea of "giving me the ring" told me, even though I didn't actually get it, that his spirit recognized, and was passing on to me, the task of a deeper spiritual exploration, perhaps outside of the church. Even though I heard he had let the Lutheran minister baptize him just before he died of emphysema, there was still **more** to know than just what the church was able to teach us, and it was OK to explore it, the dream message suggested.

In June of 1977, I was seeing a student counselor in a suburban mental health center, who needed some volunteer clients to do her internship. By coincidence she was doing an Adlerian Life Style Analysis on me for a class, and I myself had been thinking of attending the Adlerian Institute, because Robert L. Moore had recommended it for graduate work.

I would talk about how Elisabeth had inspired me, and this counselor, Joan Healy, did me the favor of saying that if I wanted to develop a closer relationship with Elisabeth, that I should be more assertive, and go for it! And, so I did!

I was just publishing my first 200 copies of *LAPIS*-- a Jungian, poetry, and spiritual exploration journal I had decided to publish, and had mailed Elisabeth a copy of the first issue. I called her up June 3rd to tell her to look for it in the mail, and just happened to catch her as she came in from gardening for a drink of water. She said, "You must be psychic," because I had timed my call so well to catch her.

I also told her that I hoped to begin painting again, but could get no vision for it.

Later this day, I took a nap, and received a dream vision. I saw a person wearing a light violet robe standing on a walled cliff or shore, looking out to sea where the sun is setting, and the light is white, pink, and gold. (It reminded me somewhat of the Two of Wands in the Tarot deck.) Then there was a scene where I am painting a white line on white canvas in a diagonal line near one of the canvas corners. I wake up, feeling peaceful.

I soon started this painting, and planned on giving it to Elisabeth when I got to see her. I called it **LAPIS**, or **LAPIS**

ONE. It began a series of paintings with LAPIS themes. The LAPIS meant to me the "Philosophers' Stone" which Carl Jung often wrote about. (Strangely enough, Jung also wrote somewhere about a 15th Century Monk and Alchemist named Degenhardus, similar to my last name of Degenhart.)

Close-up of the Lapis One painting, showing two figures, representing Elisabeth and her Egyptian Spirit Teacher. See end of chapter for full photo.

Elisabeth said that she would be in California the weekend of July 10th, to visit her "spiritual teachers" for her July 8th birthday, so I had to call her several times, and managed to arrange for a July 27th, 8 p.m. meeting.

Meanwhile, during these weeks leading up to my meeting with her, we had very unusually powerful energies come through at the Spiritualist Church that I attended at the home of Duane Bordner. July 17, I showed everyone my **LAPIS** painting that I was going to give to Elisabeth. On July 24th, the energy was so strong that everyone commented on the Entity that had come through during the sermon. The message was psychological, and something like Elisabeth would say, about how we should share our struggles with each other more. That evening, I had my friend Mark Valco take my picture with the **LAPIS** painting that I was to give to Elisabeth. [See end of chapter for that photo.]

July 24th, I also received a psychic reading at church that answered my inner question about studying to become a therapist. The psychic minister said, "You are questioning about something, and the answer is: 'It should be.'"

On July 26, 1977, the day before my meeting with Elisabeth, I had this dream: "A huge storm or tornado hits Chicago or some campground near Lake Michigan. It is caused by the spirit of an old Indian who is angry at our destruction of the land. But he spares good people and little children. Some of my books got wet in the storm."

Later the same night I also had the dream I mentioned previously about someone having the vision of building pyramids at certain energy centers all over the world to prevent earthquakes. The whole country is covered with buildings and pyramids at these special places.

I later learned that this dream occurred **one day after** Elisabeth had put down money to buy the Shanti Nilaya land, where she hoped to build a pyramid for healing purposes. I was to meet with her the very next day. Part of her vision for Shanti Nilaya at that time, was to have such healing centers all over the country! She also planned on having workshops for dying people and cancer patients at these centers. Later, her workshops could be for anyone needing to vent some feelings that needed to come out, regarding pain and trauma. Elisabeth told me on the 27th that when she shared at a lecture her idea for creating Shanti Nilaya, everyone cheered. It would also be organized like a non-profit organization, so the money involved wouldn't go to any one person.

The morning that I'm to meet with Elisabeth, I had a dream: "I am talking with my boss. He seems to think I am not satisfied with the job. But, he says encouragingly, "There is a way to serve God and receive a check for it." As if to confirm this idea, that evening Elisabeth told me that maybe someday, after I have my Masters Degree in Counseling, I could work for one of the centers that she will be setting up.

"Synchronistically, the same evening I am to meet with Elisabeth, there is a special on TV about the King Tut exhibit, which is now visiting the Chicago Museum of Natural History. I had to miss part of the show in order to leave to see her. I had been talking with a friend about going to see it, but people have had to wait for hours, or even overnight, in order to get tickets. We were thinking of going next Saturday." (From my journal.)

Then the real story begins, and much of this is taken word-for word from an account that I kept in my journal, and later typed up in August of '79:

"Elisabeth Kubler-Ross lives in Flossmoor in a Frank Lloyd Wright house, which happens to be only about two miles from where I live in Hazel Crest, Illinois. When I get there, the place seems deserted. There are no cars visible, and no one answers my knock or ring at the back door. I suddenly remember a dream I had a day or so ago in which I arrived to visit Elisabeth, and she was late by about 45 minutes because of some emergency she had to attend to. I think maybe she had gone out, and will soon be back. I sat in my car for a few minutes, waiting. Then knocked again. Finally I heard some music coming from around the corner of the house. I followed the music and came upon a sliding glass door with an indoor pool behind it. I knocked and a man came out of the pool and opened the door. I asked if Elisabeth was home, and he sort of glares at me, and the Jewish letter he is wearing around his neck also seems to glow at me, warningly. This is her husband. He asks if Elisabeth is expecting me, and I say I have an appointment. He goes to get her, while I go around to the regular back door. She has fallen asleep while watching the King Tut exhibit on TV. She had been up until four am responding to her tremendous mail."

"Two dogs meet me at the door -- a cute, white friendly one, and a dark, not-so-cute, mean-spirited one. She locks up the dark one, and the white one follows us all around. She gets some tea and some California fruit, as I inspect her house and peek into her office and library."

"Then, when she comes to sit down by me, the first thing she says is: "Are you moving?" I say that I hoped my visit with her would help me to decide, somewhat, about that. Should I move to Chicago, or stay in the suburbs, where there's grass, and commute? Commuting is kind of meditative, on the train. By the time I was done talking about moving, I had convinced myself to stay around here a bit longer. (I couldn't really afford to move out from my parents yet, anyway, and still go to school, too, taking some part time classes at the Adler Institute downtown.)"

"Then I presented my LAPIS painting to her. It was wrapped in paper with little bunny rabbits all over it. She likes bunny rabbits, and said, "Oh, my favorite paper!" I later re-read in her book, *Death, The Final Stage of Growth*, how when she was young she would hide in the woods to get away from people, and the rabbits were there to keep her company."

"Elisabeth opens the birthday card, and it happened to have a butterfly on it. This was just a "coincidence" as I had it around the house, and didn't realize until later that the **butterfly** is her symbol for Shanti Nilaya and the freedom of the spirit."

"As Elisabeth is opening the painting carefully, I say, 'This painting had to be inspired because it is the first thing I have painted in a long time. The two people in it started out to be me and you, but somehow they have changed into you and an Egyptian Teacher.' She sits the painting on the couch, trying to keep the white dog off the nice couch, (as it will do anything to be near her), when suddenly, as I am staring at it, it seems like a bolt of Light comes flashing out of the painting, from the LAPIS "Diamond" in it!! Elisabeth jumps, and gets very excited. She tells me that this is a picture of exactly what happened at her very special birthday party on Sunday (which was held on July 10th, the exact day I started the painting.) She says that her Spirit Teachers **appeared** and gave her a party! They played the flute, and she drank wine with them, just like the scene in the painting I made! (This painting had two figures in it, much like the Two of Cups Tarot card, holding cups in their hands.) She said she wore a long dress with red and white colors in it, similar to the long dress in the painting. One of her Teachers is Egyptian, just like I had painted! And they touched cups, just like in my painting!"

Elisabeth then dashed up and into a room, and brought out the dress she had worn, which I believe was red and white, and long, as in the painting (although the one in the painting was a solid orange with a red sash -- its easier to paint a solid color.) Next, she came back and said something about a **gift** they had given her -- a White Robe given her by her materialized Spirit Teachers. So, she ran back into the room, and returned with this long, white robe which had gold braid at the sleeves and neck. I reached out to touch it, and she almost stopped me. It was charged with vibrations, and I felt a slight tingle to touch it. However, my skeptical mind was still at work, too, for I thought to myself, "How can Spirits sew?" For it looked like someone had made it, and out of polyester. But this all happened so fast, I hardly had a chance to inspect the sleeve before she ran back into the room with it, saying that no one else had seen it yet, and "It's a relic."

I next showed her some poems, and an article on poetry

therapy, I had written, and gave her copies of them. The poems entitled "Illumination" and "Epitaph" were written in high school, and later published in my book *Dark Night Where Angels Tread.* (I have a few copies of this book left, which I printed, and also by 2010, it is on CafePress.com.) I also showed her some Rosicrucian Fellowship material, from the group at Oceanside, California. They said that the body should be frozen for 3 ½ days, then cremated. Elisabeth said that would be ideal, but is not necessary. Elisabeth read the poem, Illumination, and squinted her eyes, as if discovering something amazing, and said, "This one's good." It's about the crucifixion, mastering matter, and giving one's self to God. She asked if she could use my poem, "Illumination" in her workshops, and I said it was OK as long as I was given credit.

Then Elisabeth asked how long it would take to get my degree. I said "Two years." She said, "That's about the right timing; in two years we should have something set up." Then she told me about just having put the money down on the land in California to build a pyramid healing center, and that a "church" was being organized so that no one person would get the money. She would need a huge amount of money to do all this, she said, but "Miracles" would occur to make it happen. That's when I remembered the dream from the day before about the pyramids being built all over to harmonize the vibrations.

When I mentioned wanting to do things like poetry therapy, art therapy, and so on, Elisabeth suggested that I could work for one of the centers, which would be set up all over. One was being planned for Appleton, Wisconsin, at the time. As she spoke, I could suddenly see the place in my mind, and it looked like a monastery. I asked her if it was a monastery, and she said yes.

Elisabeth then mentioned a set of books that, at that time, people only heard of word-of-mouth, called *A Course in Miracles,* and asked if I had heard of it. I hadn't yet, but "coincidently" two days later an ARE catalog that came in the mail carried it, so I ordered a set, though it was a lot of money for me, then.

I told Elisabeth that I had ordered come copies of *The Quiet Mind* by White Eagle, but they hadn't come yet, so she went into a closet in her office and brought out two copies of it for me. To me, these were like "relics," having come from her as gifts, so I gave one copy to my mentor Robert L. Moore, Ph.D., who was another important influence on my life. I still have the other copy. (Even in 2010, it is here somewhere.)

As Elisabeth went into the closet, I was inspecting all her books. She has everything from *Dianetics* to Alice Bailey's entire set on *Esoteric Healing,* and she said that's only about 1/3 of all the books she has. There's more upstairs! Kubler-Ross had more occult books than Duane Bordner and I put together! I asked her how long she's been studying all this, and she responded, "Thousands of years." I knew that in public she was still being careful to avoid any connection with occultism.

I mentioned that "strange things" were happening lately, such as the strong energy we all felt at the Spiritualistic Church I attended. Elisabeth asked, "Where do you go to church?" And I explained that it was a small group in a home, where up to ten people got together and studied the Bible, esoteric interpretations and symbolism, including Qabalah, meditation, healing, and so on. She wanted to know if we studied Jung, too. I said that **I** studied Jung myself, more than the group did, and added, "I've been studying there for years." (Since I was sixteen.)

"Then she showed me an Ankh she wears around her neck which was a gift from her Egyptian Teacher named Ankah. I think of what Spiritualists call "apports," and figure the Ankh is one, if it is from a Spirit Teacher. (Apports materialize from thin air.) She mentioned the beautiful King Tut exhibit which she saw the other day, but she only had a half an hour to see it, as they waited so long, and then had to leave by a certain time. People were pushing -- it was terrible -- but so beautiful. I mentioned that I was thinking of going, if I could somehow beat the long lines."

I had also brought Elisabeth a copy of the book *The Dear And Glorious Physician,* by Taylor Caldwell, which is about St. Luke, and had a lot of Ankh symbolism in it. Believers in the "One God" of Akhetaton often used the Ankh symbol, and were the precursors to esoteric or true Christianity. The book, *Secret Places of the Lion* talks about the Ankh as a symbol used among the "Goodly Company of star-born beings dedicated to the salvation of a planet!" (Pg. 49.) The Sun Disk, or Aton, was also a symbol of these spiritual warriors. "It should be mentioned here that from the word "Aton" came atonement (at-one-ment, at-tune-ment.) From the word "Amun" the world derived amen and a-mend-ment. Atonism harmonized the entire creation so that man was ONE with his Creator. **Amunism** was a corrective discipline, for it attempted to amend and yet control human behavior through false teachings." (Pg. 50, Williamson.) (The many gods approach.)

"Then I said I wore an Ankh all through high school. The Tarot cards had one on the back of the deck. I told how I gave many lectures and card readings for the "Psychic Literature" classes. Elisabeth then got excited and wanted me to do her a card reading for right NOW. (I think she meant, for our meeting.) She looks for her cards, and cannot find them. I noticed that she has a copy of the Wilhelm/Baynes *I Ching*, and said we could do that instead. Elisabeth wonders if we will have time, as her husband will come in soon, and be grouchy if he sees her playing with all this "superstitious" stuff. I say I have a way to do it quickly with three coins, three pennies will do, and it will work just as good as the sticks, if you're psychic."

"So, Elisabeth gets her *I Ching* book and throws the coins. We get No. 22: Grace, with line 4 moving, changing into No. 30, The Clinging, Fire. Grace talks about inner simplicity being better than beauty of form alone. Line four says this: "Grace or simplicity? A white horse comes as if on wings. He is not a robber. He will woo at the right time."

"An individual is in a situation in which doubts arise as to which is better -- to pursue the grace of external brilliance, or to return to simplicity. The doubt itself implies the answer. Confirmation comes from the outside; it comes like a white-winged horse. The white color indicates simplicity. At first it may be disappointing to renounce comforts that might have been obtained, yet one finds peace of mind in a true relationship with the friend who courts him. The winged horse is the symbol of the thoughts that transcend all limits of space and time."

As I read this passage aloud to Elisabeth, it was as if other presences were in the room with us, and a tremendous energy was all around us. I felt a holy quality in the moment.

No. 30, The Clinging: To cling, to be conditioned by something. To cultivate gentleness and docility.

"I do not know for sure if this reading was for me or for Elisabeth. But it seemed that she wanted a clue to the significance of our meeting. There are many implications for both of us in this message, and it does predict my clinging to Elisabeth beyond all space and time. To her, that I am not a robber (not a negative-minded reporter), but can be trusted, is also implied. Elisabeth said that she would show her 'I Ching' to her I CHING teacher in California."

She also mentioned the next workshop she would be giving in Appleton, Wisconsin, and that I should take it, and "Discover who I really am." (As her workshops then cost $300., I couldn't afford to take one until 1983, after I had graduated with my Masters Degree in Counseling, which ended up taking me seven years to complete, going to two different schools, but finally graduating with straight A's from Governor's State University in Park Forest South, Illinois.)

"Suddenly, it was over, and I knew I should leave. She walked me out to my car, and commented about the beautiful crickets chirping, and as I got in the car, Elisabeth shook my hand, saying, "I will see you again." Then she directed me as I backed out and turned around the car to drive home."

"In my drive home, and for several days afterward, the vibrations were very powerful and peaceful all around me. Her house had been so charged with energy that you couldn't be nervous talking with Dr. Ross if you wanted to. I felt like I had visited another world. It was total 'safe space.'"

Just as my more recent meeting with Robert Morning Sky seemed to activate my inner spiritual life and dreams, so did my meeting with Elisabeth years before.

Two days later on July 29th, I woke at four a.m. with the words, "Pliney the Elder" in my mind. I didn't know who Pliney the Elder was, but looked him up in our *World Book Encyclopedia*. He was a Roman statesman, lawyer, scientist, and author of many books on natural phenomenon and history, the first set of encyclopedias. He had a nephew (probably called Pliney the Younger, I think) who sent letters to the Emperor about the "Christian problem." An Edgar Cayce book, *Edgar Cayce on The Dead Sea Scrolls*, by Glen D. Kittler, (Paperback Library, N.Y. 1970. pg. 13,) talks about him . . . ." a book called *Historica Naturalis,* written by Pliney the Elder around 70 AD., a naturalist who wrote down his observations. While traveling on duty with the Roman army Pliney did comment that while in Judea he came upon an Essene Monastery on the western shores of the Dead Sea . . . ." The *World Book* also said that he had died in 79 AD in the eruption of the volcano Vesuvius in Pompeii.

After this dream, I suddenly remembered one of my first

seances with John Scudder, where a meaningful message came through one of the psychics for me. The message was from a Spiritual Teacher of mine who they said had been a Roman Senator in life! This message was very deep, and though I could not remember it much with my mind, it addressed the deep loneliness that I felt for much of my young life. It was not a superficial message, but very profound. I was age sixteen then, and admired a history teacher named Norman Stenzel, through whose influence I began to read Jung and Hermann Hesse. It seemed that I couldn't relate to very many of the other students, as not many Sophomores study Carl Jung and read Hesse! So, the message was addressing my feeling of loneliness, and sadness that I had few friends as fascinating as Mr. Stenzel was for me then. I think I had invited him to attend a seance, and was disappointed that he never came.

I don't know if my Senator Teacher was actually Pliney the Elder, but it could be. He was a writer, and that is what I always wanted to be, too. From this memory, I did begin to realize that maybe I, too, had Spirit Teachers around me, just as Elisabeth had. And, I was beginning to get in touch with them again.

That night, I called my friend Carla, and planned on going Saturday to the King Tut exhibit. We decided not to go real early or wait all night in line for tickets, but to just take it as it comes. We'll just get there at 9 a.m., and if we could not get in, we'll do something else. But, I told her I had a very excited feeling that we would get in without waiting!

July 30th, Saturday, I dreamed I received a spiritual gift from my Teachers. It was symbolized by me finding, in the dream, a black and white spotted "apport" arrowhead on Michigan Avenue. I knew this meant I would get into the King Tut exhibit like magic.

Saturday, Carla and I arrived at the Field Museum at five minutes to nine. The doors opened at nine am. We got in a strange line at a side door, since it was shorter. People started going in, and suddenly the guard realized it, and turned around the rest of the people. Carla had just gotten in, and I said, "I'm with her!" and dashed in, the last person to get in this side door who did not have the special membership required.

So, by magic, we did not have to wait at all, when some people had been waiting all night! We didn't have to wait long to get the tickets, and were the first to get into the King Tut movie, so

Carla said, "Boy, I sure do appreciate this 'special treatment,' what ever it is!"

There was still a five hour wait until we could get into the actual exhibit, so we then took the L train up to Loyola Campus, and had a Greek dinner, meditated by Mundaline and Lake Michigan, and came back at five pm, when our number was up, to see the exhibit. We didn't wait much to get in, but it was suffocatingly crowded inside the exhibit. The energy of the objects was incredible, and especially the King Tut Mask. That did not look or feel like the work of "primitive" people to me. It was other-worldly. It was an Initiation just to be that close to the energy that it gave off.

While all these sacred things were happening, I somehow knew that I could not share much of it, even with my church group, the next day.

The Sunday and Monday after the King Tut exhibit, the vibrations were still so beautiful, and it seemed at one moment there was a presence in the room. Somehow, I knew after that I had been accepted into some kind of Spiritual Brotherhood/Sisterhood. I had passed the first tests, and now I was accepted.

I remembered a message I had received from a dream on Good Friday that same year (1977), before this all happened, which was: "You will receive a beautiful spiritual gift." For me, at this time, the meeting with Elisabeth Kubler-Ross, and all the attendant miracles, were this gift. But, now, years later, I wonder if the black and white arrowhead might have had more significance than I would have suspected then. Maybe it was part of the gift? Now, I'm finding the black and white symbol connecting me to the Hopi Koshare Clown, and teachings about the opposites. The arrowhead symbol turned out to mean more, too, when years later I converted to the Native American path, and even left the church, after studying to become a minister. And, to clinch it, my Hopi/Mexican friend Mr. B. (The Corn Farmer) considers himself Koshare, and he actually gave me a gift very recently--a black and white beaded necklace, that looks very Egyptian!

And, now, somehow, my connection with Robert Morning Sky is activating the Egyptian energy connections all over again. He's studied Egyptian Hieroglyphics for 30 years! He was looking for historical and linguistic clues to the "hidden history" that the Star

Elder had given his Grandfather. Then I dreamed of an Ankh in the Canyonlands, during a weird storm. That was an important clue for me--and it brings me back to this time of my life when meeting Elisabeth Kubler Ross had given me such inspiration and purpose.

It seems, too, as if I've failed my "assignment" in some ways. It took me forever to complete my Masters Degree, for financial and emotional reasons. I've really had to fight against inner restrictions and emotional garbage that have held me back from achieving all that I wish I had. I've lost many jobs due to saying or writing something stupid. I haven't been healed enough myself yet, so I could not heal others, like I had hoped. And, I never worked for Elisabeth, after all. Somehow, it never fell into place, and Elisabeth seemed to change, so that I wasn't as sure about limiting my focus to her brand of group/workshop therapy. And, I sure didn't want to move to Virginia, after she moved her center out there, as I had just "discovered myself," and my Indian consciousness, seemed to lead me out West.

Sadly, I have recently discovered that Elisabeth has had a stroke, which has paralyzed her left side, and moved to Arizona, where Robert Morning Sky and the Terra Society are located. She is being cared for by family and friends there. Her home was torched to the ground in Virginia, and she lost everything she owned there. That probably includes the LAPIS painting I gave her years ago, and perhaps all her awards, her Sacred Pipe, and all the "relics" she had collected over the years from her Spirit Teachers. So, there may no longer be any evidence of any of my story. That is, unless one day, while she is still with us, Elisabeth will share more of her story with us all.

(Elisabeth Kubler-Ross died on August 24, 2004, three days before I moved back in with Don on the mountain.)

So, even for us Initiates, there is always the Black and the White, the pain and the joy, the failures and the successes. The Path does not take a straight and easy course. There is life and there is death . . . .

I wasn't sure if I should put this personal story about Elisabeth in my book. How does it relate to Ancient History? Would she be upset if I shared it without her permission? At one time, I had believed in getting permission from "authority" figures. But no more. Now I'm a Renegade, and I do what Spirit calls me to do --

for the personal ego of a person such as Elisabeth has dissolved into her larger service. And her service would be strengthened if more people knew of some of the spiritual experiences she's had. She needs to share them. And I need to share mine as well.

I've been saving this story for a long time, about twenty years, and feel that now is the time to reveal it. The synchronicity with the Ankh dream and Robert's interest in Egyptian things brought this all back to light. There are many Egyptian connections in this book, and there could easily be more, if I wanted to make the book longer.

Another question that may come up is, did Elisabeth really talk with Spirit Teachers who had materialized and drank wine with her? All I know is that she believed it then, that my painting had portrayed that very event, and that I had started the painting on the same day it happened in California. This is probably not the only time they have materialized for her. Can something like this happen? Spiritualists and Native Americans believe it can happen.

In my first book, *A Medicine Story: The Return To Native American Consciousness*, I quoted a letter that Elisabeth had written to me, in which she said that she was a "Pipe woman," and reminded me of how she had so many Native American objects in her Flossmoor home. In the book *Quest* (Derek Gill, Ballantine Books, N.Y. 1980,1982), she tells of dreams and intuitions which told her that she had probably been a Native American in a past life. Thus, it should be no surprise that she may have the same kinds of spiritual experiences that Native American can have. Gregory Little, in his book, *Grand Illusions*, says that the highest form of Vision an Indian can have is for the spirit to appear physically! He writes: "Medicine dreams and vision quest visions represented the highest levels of visions possible. In these higher levels, it was common for a spirit being to appear physically to the Indian while the Indian was wide awake." (Pg. 170.) I just happened to read this statement while I was having doubts about including Elisabeth's story of materialized spirit guides. After I read it, I knew that I should share this story.

The other reason I decided to include this experience, even with some last minute doubts, was another synchronicity. The morning after having gone to bed early in turmoil over the question of including this story about Elisabeth, I happened to wake up and open up to a book that was by my bed that talked about the

Egyptian Osiris cult. (This book is *Grand Illusions* by Dr. Gregory L. Little, White Buffalo Books, Memphis, TN. 1994.) I happened to read about those who sought Initiation in this Egyptian Mystery School: "Those who were accepted faced decades of arduous tasks including mundane work as well as study and prayer ritual."

So, I thought of my prior feeling of "acceptance" into a Spiritual Brother/Sisterhood, years ago, and all the hard work, with years of schooling, and weird, mundane jobs, that I had endured.

The mystery of death and resurrection was portrayed for those seeking Initiation by red and black pillars outside of the Temple of Osiris. The opposites, again.

"At this point, the seeker was warned about the long, hard, and dangerous path he was about to embark upon. All who seek here, he was told, do not return. Some die. Others find madness. But this was the point of no return. For once the seeker entered as an initiate, he could never return to "the outside world." (Pg. 153, 154.)

"After many years of dedication, they become candidates for initiation." (Pg. 154.) They are warned of the dangers, but must overcome their fears. They are taken back to the door of the Temple of Osiris' secret sanctuary. There are tests in this secret sanctuary that could kill them. They had to pass temptations of fear, of the flesh, of understanding the "22 sacred paintings" in a great hall, "each symbolizing the Mysteries." (Pg. 154.) (Probably akin to the Tarot Major Arcana.)

The final Initiation was to go into the King's Chamber in the Great Pyramid, and be sealed into the sarcophagus, where they must die and be resurrected! Gregory Little, and the sources he is quoting, feel that the Initiate was allowed to pass out from suffocation just long enough to have a Near Death Experience, and then be resuscitated. This would enable them to "bathe in the light of Osiris." (Pg. 155.) This would give the Initiate an experience of total love and ultimate truth. The seeker becomes the Initiate, and "one of us," a full member of the Brotherhood. He has died and been resurrected.

When I read in Gregory Little's book that the Quester seeking Initiation would come to a point of no return, I suddenly remembered using that phrase at the beginning of this chapter, yet when I first wrote it, I wasn't consciously sure why I used it.

Reading his book, I knew that I had to include this chapter about Elisabeth, and that it did all fit, after all. For my writings are all about Awakening and Initiation. Meeting Elisabeth was a kind of Initiation, an acceptance into a process. Meeting Robert Morning Sky awakened that process again, and gave me a point to reflect and evaluate where I have come since twenty years ago. In my self-evaluation, I find both the black and the white, things I've done right and things done not so well. But, I must accept those things, and move on, or I will be stuck.

The other issue is that since our understanding of the true story of ancient history has been veiled deliberately from us, it is truly an Initiation to become awakened to these truths once again. There are Spiritual truths and values implied by the information as it is revealed. It is a War of Consciousness that we are in. And a War of Information. If we wish to have our spiritual freedom, we must be willing to face both the dark and the light sides of the Truth, and of ourselves. Some who seek this information die (perhaps at the hands of the Secret Government), some find madness, but for all, it's a point of no return.

Karen Degenhart holding her "Lapis One" painting shortly before presenting it to Elisabeth Kubler- Ross. (Photo by Mark Valco, 1977)

Elisabeth Kubler-Ross, MD., 1983.
Photo (c) Karen Degenhart

Karen with Elisabeth Kubler-Ross, MD. 1983.
Elisabeth interprets Karen's drawing

CHAPTER SIXTEEN:

"RELEASING THE EAGLES:"

THE TRUE MEANING OF THE "FOURTH COYOTE"

The Fourth Coyote has spoken. When I started this book I called it The Fourth Coyote because of my search for a complete Medicine Wheel made of coyote fetishes. This symbolized for me the search for completion, and a search for the missing pieces of ancient history. I have found many clues on this journey, and feel that I am on the right track. But the details still need to be discovered and organized. We each need to do much of that work for ourselves, because it is a process of consciousness and awakening. Facts and information help, but we must understand what it all means, for ourselves, and for the planet Earth.

Synchronistically, as I complete this book I happened to read in *The Voice of the Great Spirit: Prophecies Of The Hopi Indians*, By Rudolf Kaiser, (Shambhala, Boston, 1991) that those of the "Coyote clan will be a bond between the Pahana (white people) and the Hopi people." (Pg. 106, Kaiser, 1991.) A Hopi woman named Polingaysi, who belongs to the Coyote clan, believes that the whites and Hopi should work together more. My original Vision Dream was of a Marriage between an Indian man and a mostly white woman. And now my book that helps to reveal clues about ancient history shares information given by a Hopi/Apache man, Robert Morning Sky, and a Hopi/Mexican man, Mr. B. (The Corn Farmer.) Both of them have white women for companions. Coincidence? And my book is named after the Coyote. Perhaps I'm spiritually a member of the Coyote Clan. Perhaps in some other life time I believed in keeping secrets, but no more. Now is the time to at least try to Initiate anyone who is ready and open to the "dark and the light" of the process. The Fate of the Earth is at stake. (8/14/1995)

After writing the previous few pages, I thought that this book was done. However, within a couple weeks, some information came to light that felt like it needed to be included. I had experienced a kind of "group healing and channeling session" with

Barbara Barnett of Crestone, Colorado. It seemed to have released some energy blocks in me.

The morning after that session, I had a dream in which the words "Sending up the Eagles," or "Releasing the Eagles," came to me. I thought about this, and it came to me that I had read about a Hopi ceremony in which eagles were killed in order for their spirits to take the prayers of the people up to the heavens.

My memory was also jogged regarding a "drama" I use to act out when I was very young, a drama about a **dying eagle**. I use to take a blanket and put it over my shoulders, and "caw caw" as an eagle, and then die, like a martyr, a heroic victim. I was maybe five or six at the time of this "game." It was very emotional for me.

Then I thought of the Eagle Dancer, a popular Kachina statue one can buy. And I thought of the dream which said that I had "relatives in Arizona," from another life. And now I have connections with two Hopi Indian men, who I am learning from. Coincidence?

And, so I felt that part of my search for ancient history, was also a search for knowledge I once knew, in other lifetimes. So, this story is a personal journey, more so than a statement of truth. The information discovered is just part of the journey. The "Medicine Story" which started in my first book, is not yet complete. Because it is not just a journey of discovery and healing; it is a journey to find my true home.

So I pulled a book from my shelf, which, as usual, I had only glanced at previously, and looked for information about the Sacrificing of the Eagles. *The Book Of The Hopi*, by Frank Waters. (Penguin Books, N.Y. 1977) says that the eagles "are "sent home" with the kachinas." (Pg. 208.) They are killed as gently as possible. The eagle feathers are used "to carry our prayerful messages aloft to other planets and stars where life exists . . . ." (Pg. 209.) Another page (71) says, "The real Kachinas, as we know, are spirits from other planets and stars, but the high mountain to which messages are directed to them is San Francisco Mountain, Southwest of Oraibi, near Flagstaff."

Then, I flipped through this book some more and came across an amazing synchronicity. There are drawings of coyotes found at Oraibi, which indicate that "the Coyote Clan had completed all four rounds of its migration, the tail pointing to the center meaning that the clan had finally reached its permanent settlement." (Pg. 106.) Finally reaching ones permanent settlement is like arriving home after having been on ones migrational journey. This journey may have taken lifetimes, and could have traversed other planets and dimensions, as well as times and places on Earth. We are all on the Journey Home now, and the meaning of the Coyote Clan is that they are the last to arrive. When the Coyote Clan arrives home, it is the end of an Age.

Thus, the meaning of my title, unbeknownst to me when it "came" to me, suggests that the arrival of the Fourth Coyote home to the clan, signals the end of the Migrations.

Another coincidence came when I read that a Coyote drawing found at Ventana Cave, Arizona, has a "four pointed star" in front of it. Frank Waters says (Pg. 106) that the "four pointed star" or "hurry star" "warned the people they must hurry to cover the four directional legs of their migration." Apparently, the Coyote Clan would perform the duty of running ahead of the others to find a place to settle, and the four pointed star urged them to hurry. I did not know any of this when I picked the title of this book, or even suggested earlier that perhaps I am spiritually, or was once, a member of the Coyote Clan!

Yet, there is more. The Medicine Shield design that I used on my Medicine Story book, is a four pointed star! It's hoop is also striped, like the Hopi Koshare. I have been following the "four pointed star" all this time, feeling the urgency to complete the Migrations, or at least the migration of my Spirit, and then be allowed to "go home," back to the Stars . . . . The Fourth Coyote has completed a cycle on the way "home."(August 28, 1995)

In 2010, Karen lives again on the mountain in the home of Walter Don Devore, Jr., and has lived here again since September 2004. And yet, this still does not feel like her permanent "home." Perhaps there will still be more journeys or "migrations" for the Coyote before she finally completes her life journey...and finds her true home. (2/25/2010)

Karen, present day in Idaho Springs, Colorado.

KAREN DEGENHART
BOOK ORDERING INFORMATION

First check http://ebay.com
Seller Starwoman77

A MEDICINE STORY: The Return To Native American Consciousness
http://karendegenhart.com

BEAUTIFUL SIMPLICITY: The Journey Into A Hidden Reality
http://cafepress.com/BeautSimplicity (or)
http://cafepress.com/BeauSimplicityY

THE FOURTH COYOTE: A Personal Search For the Hidden History of Earth
http://createspace.com/3441391

DARK NIGHT WHERE ANGELS TREAD
http://cafepress.com/DarkNightAngels

THE BIRTH OF A POET: Blossoms of Faith, and Dark Journey,
http://cafepress.com/BirthofaPoet

WOUNDED SHAMAN
http://cafepress.com/WoundedShaman (or)
http://createspace.com/3332323

Thunderbird Publications
Karen Degenhart
480 Overlooked Way
Idaho Springs, CO. 80452.
http://karendegenhart.com
kdegenhart8@gmail.com

Made in United States
Orlando, FL
11 November 2026

90609391R00085